Chaos Monitoring in Dynamic Systems - Analysis and Applications

Edited by Louay S. Yousuf

Published in London, United Kingdom

Chaos Monitoring in Dynamic Systems - Analysis and Applications
http://dx.doi.org/10.5772/intechopen.100745
Edited by Louay S. Yousuf

Contributors
Juan Carlos Jauregui-Correa, Emarti Kumari, Yongjian Sun, Zihan Wang, Jesús Alan Calderón Chavarri, Eliseo Benjamín Barriga Gamarra, Julio César Tafur Sotelo, Rodrigo Alonso Urbizagástegui Tena, Dante Jim Randal Gallo Torres, Jaime Eduardo Zeña Delgado, Christian Enrique Gózar Pastor, John Hugo Lozano Jáuregui, Mauricio A. Ribeiro, Jose M. Balthazar, Ângelo M. Marcelo Tusset, Átila M. Bueno, Hilson H. Daum, Louay S. Yousuf

First published in London, United Kingdom, 2024 by IntechOpen
IntechOpen is the global imprint of INTECHOPEN LIMITED, registered in England and Wales, registration number: 11086078, 5 Princes Gate Court, London, SW7 2QJ, United Kingdom

British Library Cataloguing-in-Publication Data
A catalogue record for this book is available from the British Library

Additional hard and PDF copies can be obtained from orders@intechopen.com

Chaos Monitoring in Dynamic Systems - Analysis and Applications
Edited by Louay S. Yousuf
p. cm.
Print ISBN 978-1-80356-434-0
Online ISBN 978-1-80356-435-7
eBook (PDF) ISBN 978-1-80356-436-4

Meet the editor

Dr. Louay S. Yousuf is a lecturer professor in the Department of Mechanical Engineering, San Diego State University, USA. He received his Ph.D. from the University of Baghdad, Iraq, and was a postdoctoral fellow at the Mechanical Engineering Department, University of Auburn, USA, before joining San Diego State University. His research interests include nonlinear dynamics, composite materials, finite element methods, mechanical vibrations, mechanisms, and robotics.

Contents

Preface

Chaotic phenomenon has been and continues to be an essential component in the nonlinear vibration of dynamical systems at different degrees of freedom. The nonlinear response of vibration is employed to describe the movement of the dynamical system. This book discusses the chaotic phenomenon in cam-follower systems, rolling bearings, high-rise buildings, and vehicle track interaction. The experimental data included supports the numerical simulation results. The Lyapunov exponent parameter is used to quantify the level of chaos in nonlinear dynamics systems. Moreover, the Fast Fourier Transform (FFT), bifurcation diagram, Poincaré map, and phase-plane diagram are used in nonlinear dynamics behaviors. SolidWorks, MEMS, and ANSYS programs are used to detect the nonlinear response of dynamical systems in different applications of sciences.

Louay S. Yousuf
Mechanical Engineering Department,
San Diego State University,
San Diego, California, USA

Chapter 1

Effect of Different Material Properties on the Nonlinear Dynamics Phenomenon of Cam-Follower Mechanism

Louay S. Yousuf

Abstract

The effect of different material properties for both radial cam and translated roller follower on the nonlinear dynamics phenomenon is investigated at different cam speeds. The equations of the dynamic motion of the follower movement are derived based on the conception of mechanical vibration theory. In this chapter, we investigate the nonlinear dynamics behavior by detecting it using phase-plane diagram and Poincare' map at different material properties and different cam speeds of the cam. The follower is moved with three degrees of freedom inside its guides and the cam is spinning about a fixed pivot. The nonlinear response of the follower is calculated using SolidWorks program at different cam speeds, different guides' clearances, and different material properties. The nylon, acrylic, and aluminum greasy and steel dry and greasy material properties are examined for nonlinear dynamics behavior of cam-follower mechanism.

Keywords: phase-plane diagram, Poincare' map, radial cam, translated follower, different material properties, nonlinear dynamics, non-periodic motion

1. Introduction

The radial cam and translated roller follower mechanism can be found in windshield wiper of the car in which the oscillating motion of the wiper is controlled by the rotary motion of the cam. This kind of cam has been selected based on the irregular profile in which it has two noses and two flanks. At high speeds of the cam, the follower will jump off the cam and introduce chaotic motion at the locations of the tip of the noses. The application of the proposed mechanism is in controlling the opening and closing of valves in internal combustion engine. Nonlinear dynamics phenomenon has happened in the presence of the follower guide's clearance. The mathematical model of transient impact, separation, and contact is described by Yang et al. using an oblique impact in cam-follower system [1]. They explained that at low speed for the cam and without the use of coefficient of restitution, the cam, and the follower are in permanent contact. Due to the nonlinear dynamics phenomenon, the largest Lyapunov exponent, the power spectrum of Fast Fourier Transform (FFT), Poincare' map are discussed by Yousuf to investigate the detachment between the cam and the follower. The response of the follower is calculated at different coefficients of restitution, different cam speeds, and different follower Guides' clearaances [2]. By taking into consideration the coefficient of restitution, Sundar et al. analyzed the model of nonlinear contact damping and contact stiffness of single degree of freedom system [3]. Moreover, the same group in Ref. [3] discussed the effect of rolling and sliding contact on the nonlinear dynamics phenomenon [4]. Osorio et al. studied the bifurcation of corner impact at variable cam speeds since the detachment occurred between the cam and the follower [5]. Li and Du used the model of coefficient of restitution to control and analyze the bifurcation diagram of collision constrained vibrational system. On the other hand, phase-plane diagram has been built between the displacement and velocity of the follower due to the energy dissipation outside the phase-plane envelope, while the Poincare' map [6] shows the single and group points of black dots. The aim of this chapter is to study the effect of different contact material properties on the nonlinear dynamics phenomenon for radial cam and translated roller follower system at different cam speeds and different follower guides' clearances.

2. Phase-plane diagram and Poincare' map of nonlinear dynamics phenomenon

Four follower guides' clearances (C = 16·10-3, 17·10-3, 18·10-3, and 19·10-3 mm) at different cam speeds are used in SolidWorks program. The spring between the follower and the installation table works as a secondary force actuator. The total follower displacement is shown in the following eq. [7]:

$$X_c(t) = e^{-\beta t}\left(C_1 \mathrm{Sin}\left(\sqrt{\omega_n^2 - \beta^2}\, t\right) + C_2 \mathrm{Cos}\left(\sqrt{\omega_n^2 - \beta^2}\, t\right)\right) + \frac{X_{st}}{H}\left[1 - \left(\frac{\Omega^2}{\omega_n^2}\right)\right]\mathrm{Sin}(\Omega t) - \frac{X_{st}}{H}\frac{2\beta\Omega}{\omega_n^2}\mathrm{Cos}(\Omega t) \tag{1}$$

Where:

$$\frac{c}{m} = 2\beta \tag{2}$$

$$\frac{k}{m} = \omega_n^2 \tag{3}$$

$$P = \frac{F}{m} \tag{4}$$

$$H = \left[1 - \left(\frac{\Omega^2}{\omega_n^2}\right)\right]^2 + \frac{4\beta^2\Omega^2}{\omega_n^4} \tag{5}$$

$$X_{st} = \frac{P}{m\omega_n^2} \tag{6}$$

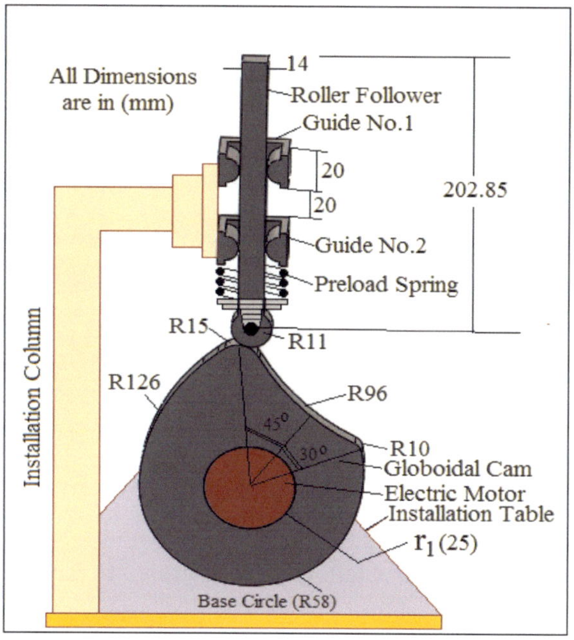

Figure 1.
Cam-follower system with its dimensions.

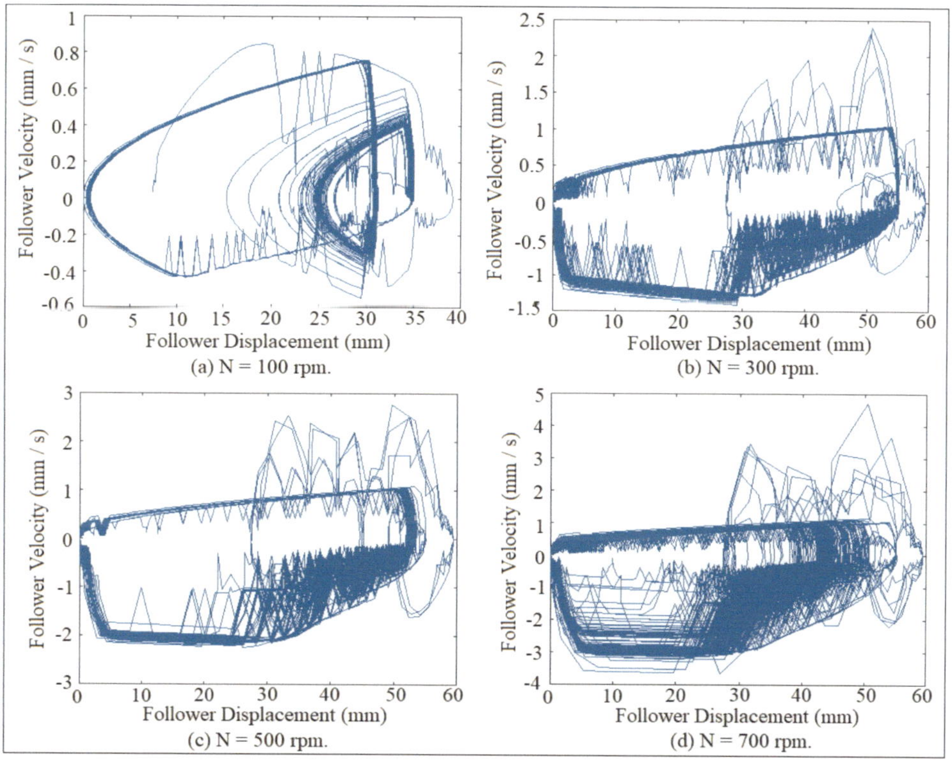

Figure 2.
Phase-plane mapping when the contact condition is aluminum greasy for follower guide's clearance (16.10-3 mm).

$$C_1 = \frac{\Omega X_{st}\left[\frac{\Omega^2}{\omega_n^2} + \frac{2\beta^2}{\omega_n^2} - 1\right]}{H\sqrt{\omega_n^2 - \beta^2}} \tag{7}$$

$$C_2 = \frac{2X_{st}\beta\Omega}{H\omega_n^2} \tag{8}$$

The preload spring is included with the contact force as in below [8] in Eqs. (9) and (10):

$$P_c = \frac{1}{Cos(\phi)}\left[K(\Delta + X_c(t)) - KX_c(t) - m\ddot{X}_c(t)\right] \tag{9}$$

$$\tan(\phi) = \frac{\dot{X}_c(t)}{X_c(t) + R_b^2} \tag{10}$$

In which $K(\Delta + X_c(t))$ is the preload spring.

SolidWorks program already has the library of material properties since both the radial cam and the translated roller follower are assumed to have the same material

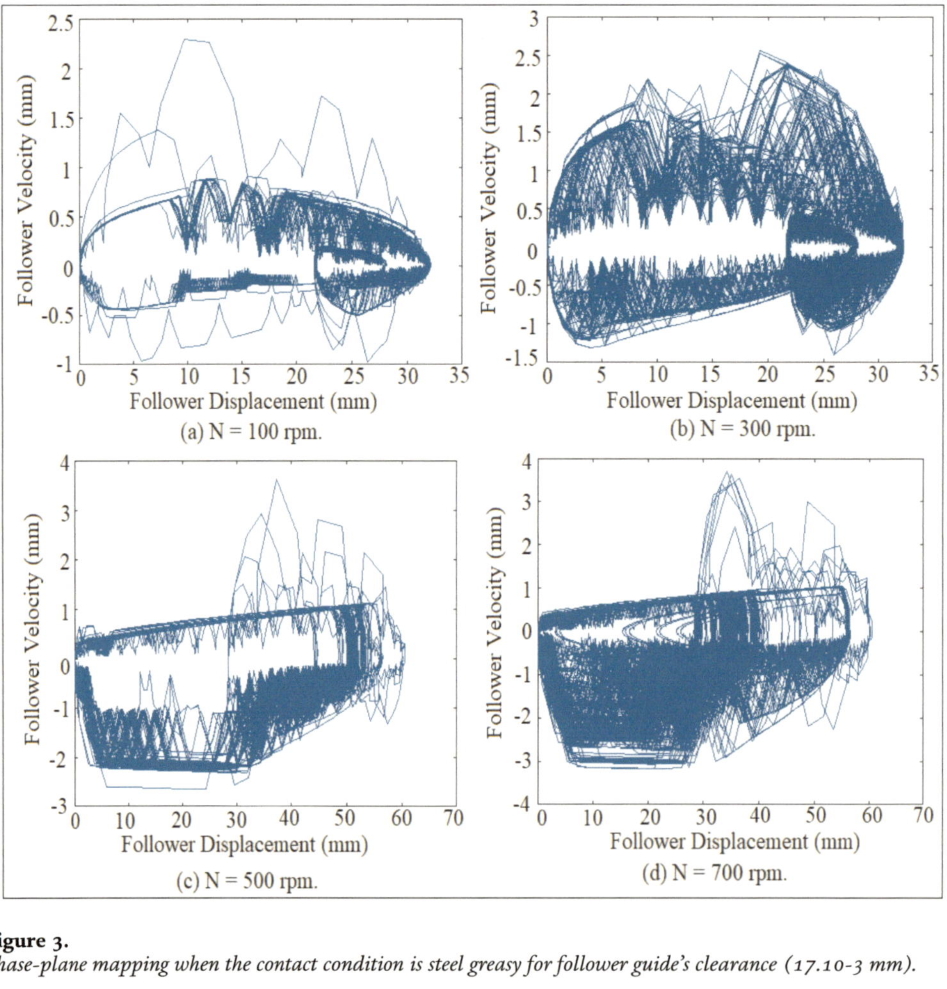

Figure 3.
Phase-plane mapping when the contact condition is steel greasy for follower guide's clearance (17.10-3 mm).

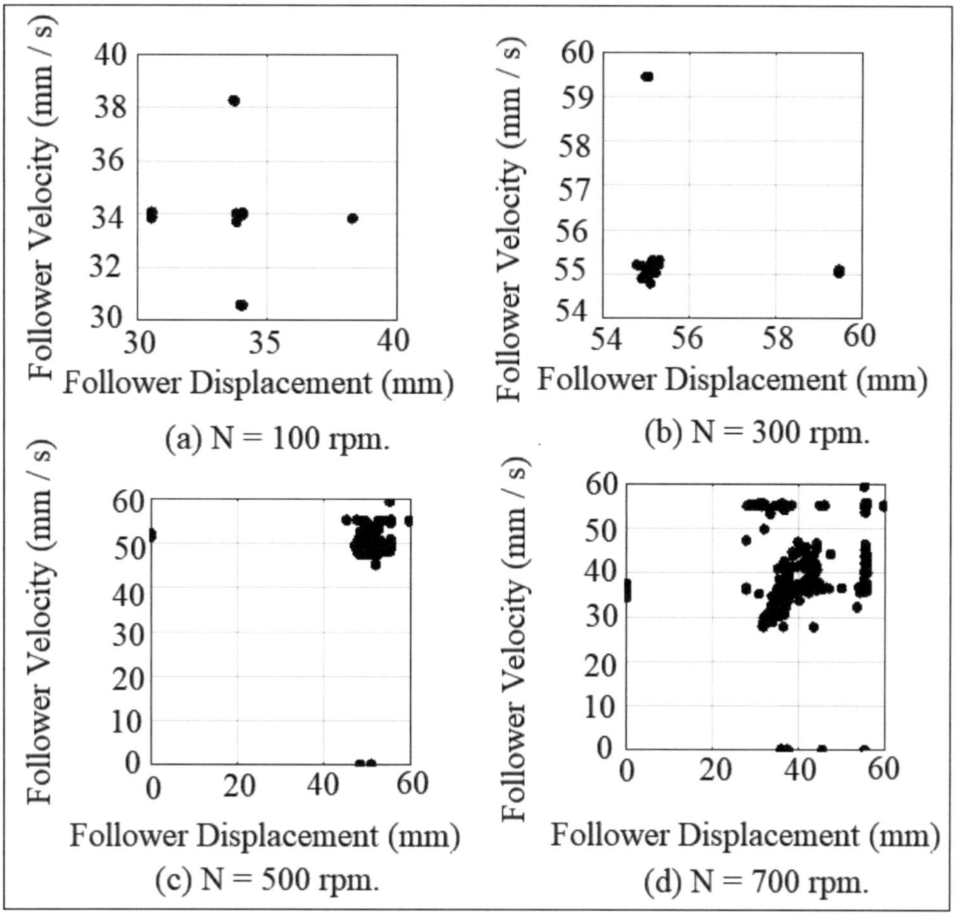

Figure 4.
Poincare' map when the contact condition is nylon for follower guide's clearance (16.10-3 mm).

properties at the contact point. Different material properties for the radial cam and translated roller follower such as steel greasy, steel dry, aluminum greasy, aluminum dry, nylon, and acrylic are considered in the contact model in SolidWorks program. In the future study, the different material properties at the contact point between the cam and the follower are taken into account. In other meaning both the cam and the follower have different material properties. **Figure 1** shows the mechanism of cam-follower with its dimensions.

Phase-plane diagram shows how the attractor of the follower displacement-velocity grows or shrinks over the time at different contact conditions. Phase-plane diagram is another proof of chaotic motion alongside with Poincare' map. The broken lines in the upper and lower surfaces in the phase-plane diagram show the effect of impact in one cycle of the cam rotation for the given follower displacement and follower velocity. The broken lines increased with the increasing of cam speeds and with the increasing of follower guides' clearances. When the orbit of the follower displacement-velocity in state space domain forms a closed cycle, it signifies periodic motion. When the attractor of the follower displacement-velocity diverges with no limit of spiral cycles, it indicates non-periodic and chaotic motion [9]. **Figure 2** shows the phase-plane mapping when the contact condition is aluminum greasy for follower guide's clearance (16.10–3 mm) at different cam speeds. The cross-linking of the follower displacement-velocity orbits increases as the cam speeds increase starting from follower displacement (30 to 50 mm) as indicated in **Figure 2b–d**. The follower motion variation is increased with the increasing of follower velocity since there will be an energy dissipation outside the envelope of phase-plane diagram. The system in **Figure 2a** shows the quasi-periodic motion for the follower displacement since the follower starts double impact and detachment at follower displacement (30 mm) and the follower comes back to the cam at follower displacement (25 mm). The motion of the follower in **Figure 2b–d** shows the non-periodic motion for the follower displacement. **Figure 3** shows the phase-plane mapping when the contact condition is steel greasy for follower guide's clearance (17.10-3 mm) at different cam speeds. The multi and double impacts have occurred when the follower displacement is between (22 and 31 mm) as indicated in **Figure 3a** and **b** at (N = 100 and 300 rpm) respectively. The cross-linking of the follower displacement is increased with the increasing of cam

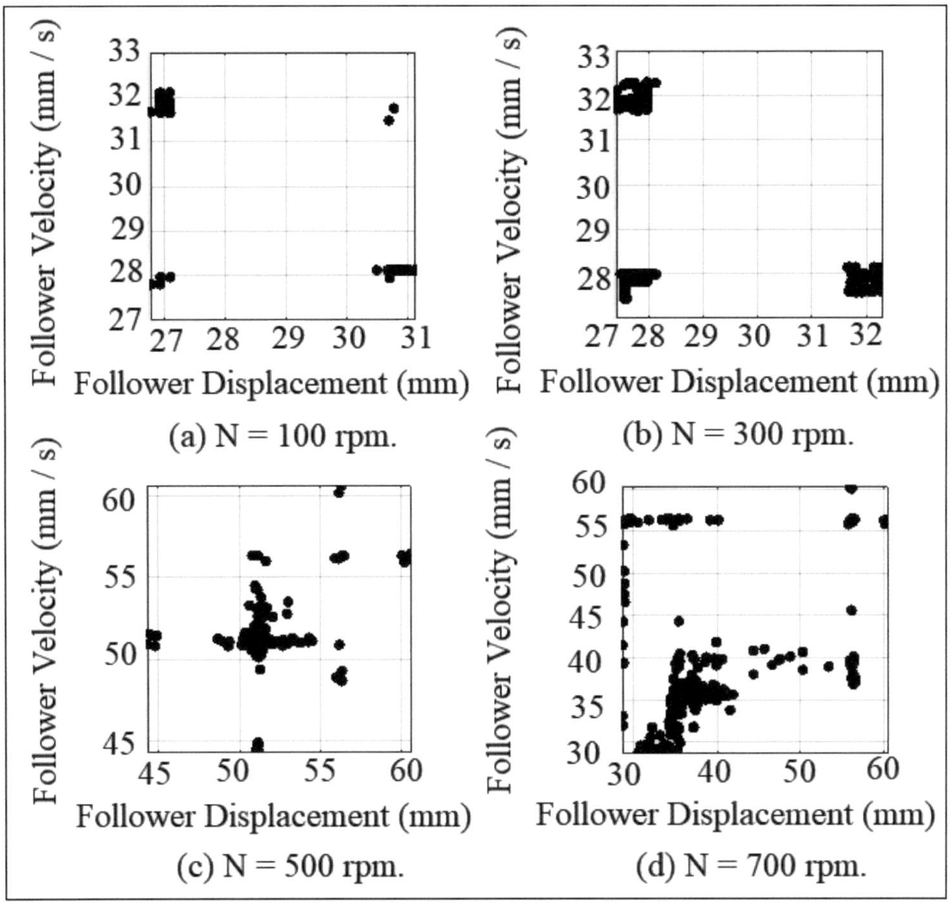

Figure 5.
Poincare' map when the contact condition is steel greasy for follower guide's clearance (17.10-3 mm).

speeds and follower guides' clearances since there will be an energy dissipation outside the envelope of phase-plane diagram. The non-periodic motion of the follower displacement is shown in **Figure 3c** and **d** when the follower starts multi-impacts since the follower variation is increased with the increasing of cam speeds. The Poincare' maps investigate the chaotic motion of the follower due to multi-impact in one cycle of the cam rotation [10]. **Figure 4** shows the Poincare' maps when the contact condition is nylon for follower guide's clearance (16.10-3 mm). Poincare' map represents that the follower motion reached the steady state (periodic motion) as shown in **Figure 4a** based on the single dot in Poincare' map for the given displacement and velocity. **Figure 4b** reflects the periodic motion of the follower displacement at (55 and 59.5 mm) while the non-periodic motion is occurred when the black dots are stationed around one area inside Poincare' map. The more black dots in Poincare' maps, the more detachment heights between the cam and the follower. The non-periodic motion of the follower displacement is indicated in **Figure 4d**. The quasi-periodic motion of the follower displacement is shown in **Figure 5a** and **b** at follower displacement (27 mm), and (30.5 to 31 mm). The non-periodic motion of the follower displacement is shown in **Figure 5d**. The black dots are increased with the increasing of cam speeds and with the increasing of follower guides' clearances. SolidWorks software is used to calculate the follower displacement and follower velocity at different contact conditions, different cam speeds, and different follower

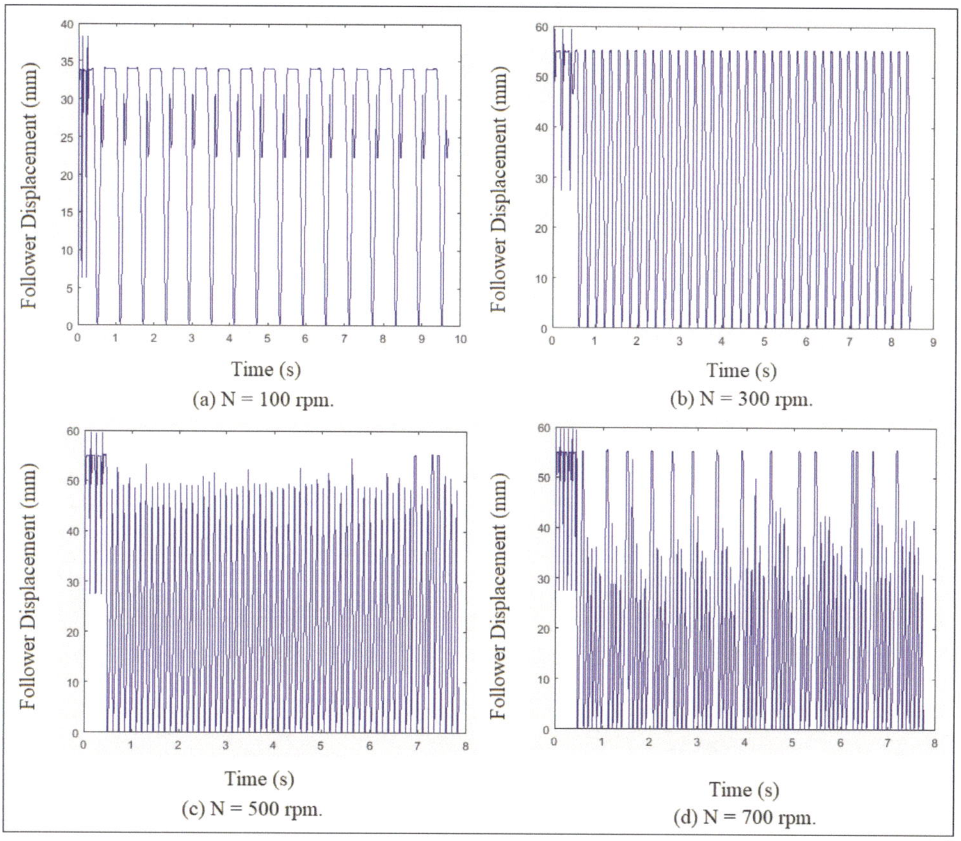

Figure 6.
Follower displacement mapping when the contact condition is nylon for follower guide's clearance (17.10-3 mm).

guides' clearances. Follower guide clearances (C = 16·10-3, 17·10-3, 18·10-3, and 19·10-3 mm) are used in the simulation while the cam is rotating at constant speed (100, 300, 500, and 700 rpm). **Figure 6** shows the mapping of follower displacement when the contact condition is nylon at different cam speeds and follower guide's clearance (17.10-3 mm). The cam and the follower are in permanent contact as shown in **Figure 6a** in which it indicates periodic motion. The quasi-periodic motion of the follower displacement is shown in **Figure 6b**. The detachment of the follower is shown in **Figure 6c** in which it indicates non-periodic motion and chaos as shown in **Figure 6d**. **Figure 7** shows the mapping of follower displacement when the contact condition is steel greasy at different cam speeds and follower guide's clearance (17.10-3 mm). The cam and the follower are in permanent contact as shown in **Figure 7a** and **b** which indicates periodic motion for the follower displacement. The quasi-periodic motion of the follower displacement is shown in **Figure 7c** in which the detachment is occurred between the cam and the follower. **Figure 7d** shows the non-periodic motion and chaos for the follower displacement due to the high speed of the cam. **Figure 6d** indicates non-periodic motion and chaos over all the periods of time

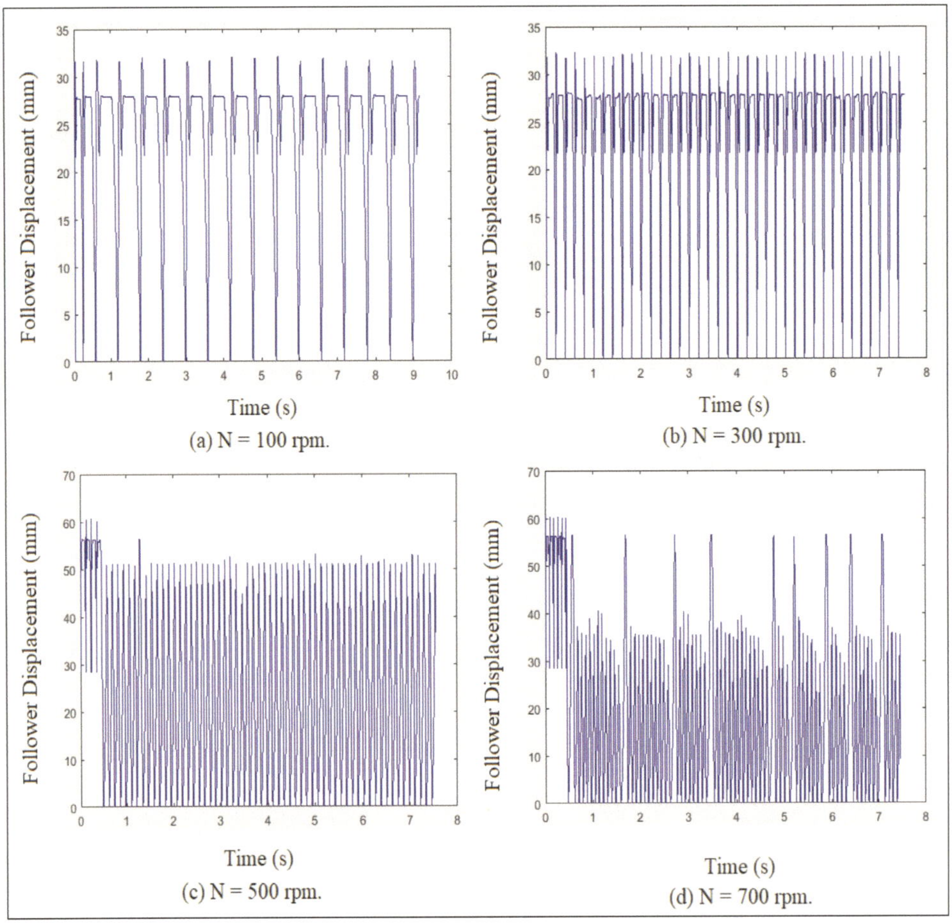

Figure 7.
Follower displacement mapping when the contact condition is steel greasy for follower guide's clearance (17.10-3 mm).

DOI: http://dx.doi.org/10.5772/intechopen.112795

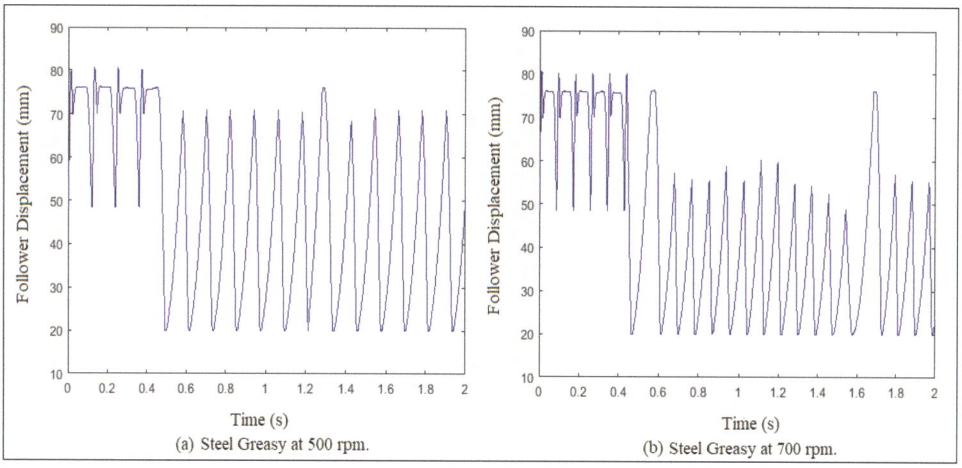

Figure 8.
Follower displacement mapping when the contact condition is steel greasy for follower guide's clearance (17.10-3 mm).

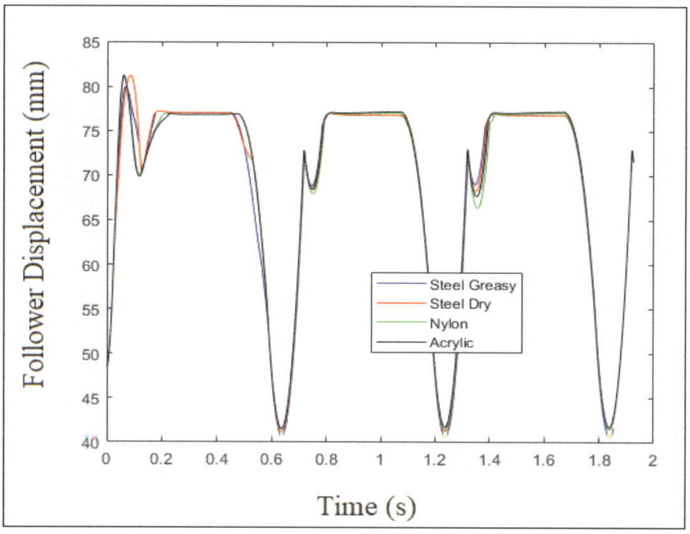

Figure 9.
Follower displacement mapping for follower guide's clearance (19.10-3 mm) at different contact conditions.

while in **Figure 7d** the motion of the follower displacement is divided between periodic and non-periodic motion and chaos over some periods of time. It can be concluded from all the contact conditions that the non-periodic motion starts at (N = 300 rpm) except for the contact condition of nylon material properties where the non-periodic motion starts sometimes earlier after (N = 100 rpm). All the contact conditions have periodic motion at (N = 100 rpm). **Figure 8** shows the follower displacement mapping when the contact condition is steel greasy for follower guide's clearance (17.10-3 mm) while **Figure 9** shows the follower displacement mapping for follower guide's clearance (19.10-3 mm) at different contact conditions. **Figure 10** shows the follower displacement mapping when the contact condition is aluminum greasy at different guides' clearances.

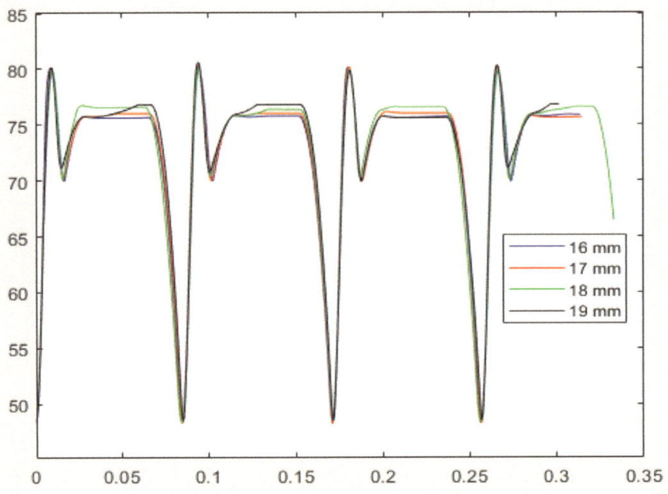

Figure 10.
Follower displacement mapping when the contact condition is aluminum greasy at different guides' clearances.

3. Conclusions

Phase-plane diagram and Poincare map are used to detect the nonlinear dynamics phenomenon in radial cam and flat-faced follower mechanism at different cam speeds and different contact condition of the material properties. As stated, the level of chaos of steel greasy material in nonlinear dynamics behavior is more than the level of chaos of nylon material properties due to the contact between the cam and the follower. The peak of rise stroke is increased for steel dry and greasy while it decreases for nylon and acrylic material properties. The dwell stroke is varied and increased with the increasing of guide's clearance of aluminum greasy material properties. The level of chaos of nonlinear dynamics of steel greasy is more than the level of chaos of aluminum greasy since this level is increased with the increasing of cam speeds.

Nomenclatures

$X_c(t), \dot{X}_c(t), \ddot{X}_c(t)$	Linear displacement, velocity, and acceleration of the roller follower, mm, mm/s, mm/s^2.
m	Mass of follower stem, kg.
Ω	Cam angular velocity, rad/s.
X_{st}	Static deflection, mm.
ω_n^2	Follower angular velocity, rad/s.
t	Time, s.
F	External force on the follower, N.
c	Damping coefficient, N·s/mm.
P_c	Contact force between the cam and the follower, N.
K	Spring stiffness, N/mm.
R_b	Radius of base circle of the cam, mm.
Δ	Preload deflection before the cam starts spinning, mm.
ϕ	Pressure angle, angle.

Author details

Louay S. Yousuf
Mechanical Engineering Department, San Diego State University, San Diego, California State, USA

*Address all correspondence to: louaysabah79@yahoo.com

References

[1] Yang YF, Lu Y, Jiang TD, Lu N. Modeling and nonlinear response of the cam-follower oblique impact system. Journal of Discrete Dynamics in Nature and Society. 2016;**2016**:1-8

[2] Yousuf LS. Detachment detection in cam follower system due to nonlinear dynamics phenomenon. Journal of Machine, Special Issue: Dynamics Analysis of Multibody Mechanical Systems. 2021;**9**(12):349

[3] Sundar S, Dreyer JT, Singh R. Rotational sliding contact dynamics in a non-linear cam-follower system as excited by a periodic motion. Journal of Sound and Vibration. 2013;**332**(18): 4280-4295

[4] Sundar S. Impact damping and friction in non-linear mechanical systems with combined rolling-sliding contact. [PhD thesis]. New York City, USA: The Ohio State University; 2014

[5] Osorio G, di Bernardo M, Santini S. Corner-impact bifurcations: A novel class of discontinuity-induced bifurcations in cam-follower systems. SIAM Journal on Applied Dynamical Systems. 2008;**7**(1):18-38

[6] Li Z, Du Y. Interval of restitution coefficient for chattering in impact damper. Journal of Low Frequency Noise, Vibration and Active Control. 2021;**2021**:1461348421105

[7] Yousuf LS. Nonlinear dynamics behavior of cam-follower system using concave curvatures profile. Journal of Advances in Mechanical Engineering. 2020;**12**(9):1687814020945920

[8] Yousuf LS. Influence of nonlinear dynamics behavior of the roller follower on the contact stress of polydyne cam profile. Journal of Processes. 2022; **10**(3):585

[9] Yousuf LS. Experimental and simulation results of a cam and flat-faced follower mechanism. Journal of Computational and Nonlinear Dynamics. 2017;**12**(6):061001

[10] Yousuf LS. Nonlinear dynamics phenomenon detection in a polydyne cam with an offset flat-faced follower mechanism using multi shocks absorbers systems. Journal of Applications in Engineering Science. 2022;**9**:10086

Chapter 2

Perspective Chapter: Optimal Analysis for the Correlation between Vibration and Temperature through an Intelligent Sensor/Transducer Based in Amorphous Nanostructures to Measure Vibrating Surfaces Temperature

Jesús Alan Calderón Chavarri, Julio César Tafur Sotelo, Eliseo Benjamín Barriga Gamarra, John Hugo Lozano Jáuregui, Dante Jim Randal Gallo Torres, Rodrigo Alonso Urbizagástegui Tena, Jaime Eduardo Zeña Delgado and Christian Enrique Gózar Pastor

Abstract

The vibration is an oscillatory movement caused by a propagation of waves through fluids or solids, and this consequence is achieved in many mechanic systems by the energy transmission between the movement source with the machine that needs the transmission movement, such as the vibration produced by a combustion engine, by a compressor system and by a result of movement transmission over rotor systems. However, if it is not a controlled mechanism to moderate the produced decibels, the main system that is affected by the vibration can reduce its performance; moreover, it can increase the surface temperature of the vibrating source and systems around. In spite of this, when it uses contact sensors to measure the vibration and temperature over the surface vibrating system, the measured data are under disturbance caused by the vibration source. Therefore, in this research is proposed an intelligent sensor/transducer based in amorphous nanostructures owing to measure the vibration of the surface through infrared (IR) emitter/receiver and the absorbance of the receiver sample has a quite range of work and robustness under disturbance of vibrating signals. This proposed sensor also has the possibility to charge energy by itself because of sun/warmth energy conversion.

Keywords: temperature measurement, vibration measurement, sensor/transducer design, mathematical modeling, wireless communication, calibration, combustion motors

1. Introduction

The measurement of vibration and temperature is quite important according to get information regarding the oscillations [1, 2] and the molecular kinetic energy of the movement source and heat sources of engines internal components, or combustion motor's external surfaces. However, there are tasks, in which is not possible to get contact between the sensor with the movement source and thermal source, hence the temperature measurement is given through IR sensors. In other side, it was possible to find the wave parameters from the IR signal of the thermal source owing to get an estimation of the vibration frequency of the movement source. Notwithstanding, there is a trouble concerning the transduction stage in the measurement while there is not a transducer algorithm designed as a consequence of mathematical model which correlates the calibration data with theoretical model of the heat transfer and the surface vibration of the movement source and thermal source. For this reason, it was proposed in this research to analyze a mathematical procedure of the measurement instrumentation according adaptive coefficients in MF strategies [3]. In this research is explained and analyzed the temperature measurement process and the transduction process as the strict correlation with the IR signal from the thermal source.

Therefore, the proposed sensor was evaluated for the measurement of the temperature and vibration of a combustion motor because of getting the understanding of its combustion and the motor user could achieve its diagnostic. There are many temperature sensors based in passive measurement such as thermocouples and thermistors, which proportionate the correlation of temperature in electrical equivalence of voltage and electrical resistance respectively, by other side, there are different piezoelectric to measure the vibration (frequency and amplitude) of surfaces. Nevertheless, there are tasks to measure vibration and temperature of systems that are located in intricate places and it is not suggested to use contact sensors [4–11], hence IR sensors are the appropriated solution. For example combustion motors have many components inside and the combustion process can produce vibrations in them and in all the motor. Moreover, a not fuel good quality can cause damage in the combustion motors which are plenty used in industry and transport such as in Peru, where the transport (public and private) depends of this kind of motor and will continue using them during many years yet, in spite of the new technology in motors are enhanced by electric motors (or hybrid), hence, it is necessary to understand the combustion process in combustion motors. Therefore, temperature and vibration sensors can give information of the operation of the combustion motor according to repair them when its components are not working appropriated and as a consequence to look for their reparation avoiding pollution. To visualize the measured information, the sensor data (electrical equivalence of temperature and vibration) needs electronic devices due to compensate, amplify and linearize the electrical equivalent of temperature transduced to thermal units. In this research is proposed a mathematical model strategy based in MF in order to get the transduction result. The complexity of the transduction is replaced by the mathematical model designed.

The proposed sensor is part of an integrated system, which is depicted by the **Figure 1**. The proposed sensor is represented as IS (Intelligent Sensor) in order to measure temperature and vibration (T&V) and the measured data can be sent through IR to an external computer that can be at many meters of distance and 2 antennas A1 and A2 have the tasks of the data transmission. Moreover, the IS has independence of its own energy to be operating because of this proposed sensor has integrated a sample of sun panel to obtain electrical energy through sun/heat energy conversion.

Hence, in this research is proposed an intelligent sensor/transducer (as part of an integrated system) based in nanostructures due to measure the temperature and vibration of the combustion motor surface depicted by **Figure 2**, in which "A" is the IR emitter in controlled frequency that could not be confused with the IR signal produced by temperature of the combustion motor operation. "B" is the sample transducer to receive the vibration signal, for which "D" is the vibration sensor based in nanotubes amorphous. "C" is the Anodic Aluminum Oxide (AAO) sample transducer to receive the temperature signal for which "E" is the temperature sensor based in nanoholes amorphous. "F" is the battery to proportionate energy of the proposed sensor, and "G" is data transmitter according to send the temperature signal and vibration signal from the surface of the combustion motor to a receptor which can be used by the user due to get the diagnostic of the combustion motor. Moreover, H is the sample based in nanostructures to receive the sun energy and I is the converter to electrical equivalent signal due to store the energy in the batteries of the sensor. Many systems can improve their monitoring variable by advanced sensors such as sensors based in nanostructures [1] thereby the understanding of the geometry and material

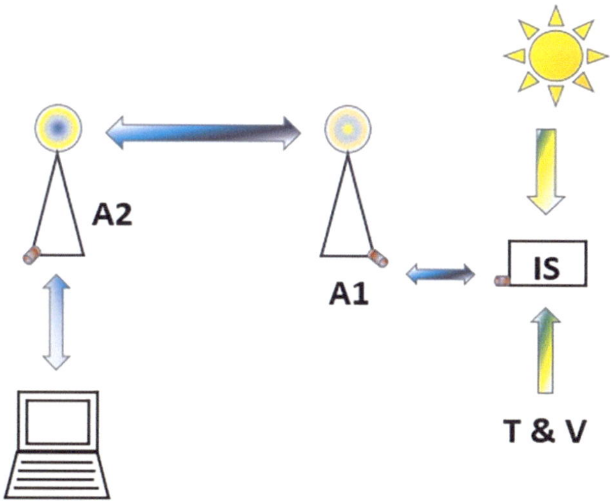

Figure 1.
Communication system of the proposed IS.

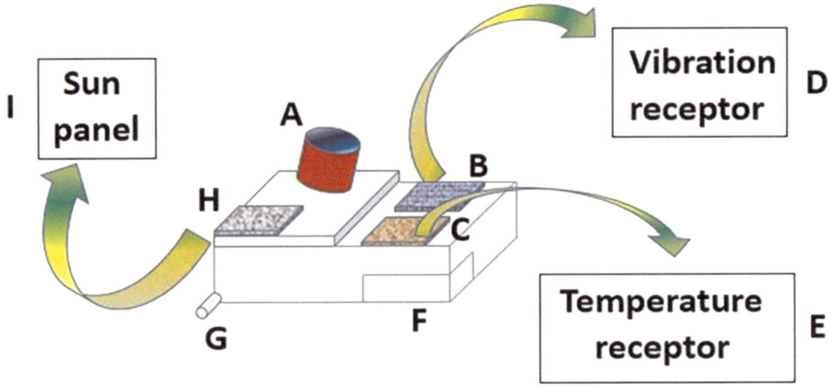

Figure 2.
Temperature and vibration transducer design.

of the sensor is quite important due to get the optimal transduction as consequence of the measurements.

2. Analysis of the optimal transduction design

Optimization analysis through MF is briefly studied in this chapter, in order to find the best solution for the data interpretation from the designed intelligent sensor, whereby the theoretical models of the physical process and analytic interpretation of the experimental results give the support to achieve the costing function by multivariable systems, because of the correlation among the internal variables of the system with the costing function.

Therefore, it was analyzed by polynomial equations as it is described by the general model in Eq. (1), for which "D^n" is the derivative $\frac{\partial^n}{\partial t^n}$, $y(t)$ and it is the variable output matrix, "$u(t)$" is the variable input matrix, "$e(t)$" is the variable error matrix, "a" and "b" are the adaptive coefficients of the system [3, 12].

$$D^n y(t) + \sum_{j=1}^{n} a_j D^{n-j} y(t) = \sum_{j=1}^{n} b_j D^{n-j} u(t) + e(t) \tag{1}$$

Where solution error analysis, "$e(t)$", is discrete error, and "V" keeps the Fourier series coefficients, which is given by the Eq. (2) [3, 12].

$$e_n(m) = \sum_{k=m}^{n+m} \alpha(k, m, \theta_a)\, V(k) \tag{2}$$

Furthermore, α is the frequency parameter function given by Eq. (3) [3, 12].

$$\alpha(k, m, \theta_a) = C_{k-m} \sum_{j=0}^{n} a_j (jkw_0)^{n-j} \tag{3}$$

For which, the nonlinear model for the error analysis is given by the Eq. (4) [3, 12].

$$\sum_{j=0}^{n_1}\sum_{k=1}^{n_2} g_j(\theta)F_{jk}(u,y)P_{jk}(p)E_k(u,y) = 0 \tag{4}$$

Therefore, the costing function is given by Eq. (5) [3, 12].

$$J(\theta) = \sum_{j=0}^{n_1}\sum_{k=0}^{n_1} r_{jk}g_j(\theta)g_k(\theta) \tag{5}$$

Also, according to get parameters of the main model, it was achieved the derivation showed by Eq. (6) [3, 12].

$$\frac{\partial J}{\partial \theta} = (\Upsilon - \Gamma\theta)^T W^{-1}(\Upsilon - \Gamma\theta) \tag{6}$$

Where parameters are showed in Eq. (7), as the dependence on the adaptive coefficients, in which Υ is the response matrix, Γ is the internal variables matrix, W is the weight matrix and θ is the sensor parameters matrix [3, 12].

$$\theta = \left(\Gamma^T W^{-1}\Gamma\right)^{-1}\Gamma^T W^{-1}\Upsilon \tag{7}$$

The interpretation and scheme for optimization is depicted by **Figure 3**, in which are presented 3 paths C1 (green color curve), C2 (violet color curve) and C3 (red color curve) according to achieve the position B from the position A. C2 represents the theoretical path such as the theoretical variable of a process, C3 represents the experimental path due to an experimental data, therefore C1 is the optimal path owing to achieve the position B.

3. Modeling

It is necessary to design the mathematical model of the proposed sensor that is possible to do through the interpretation of the problematic described in chapters

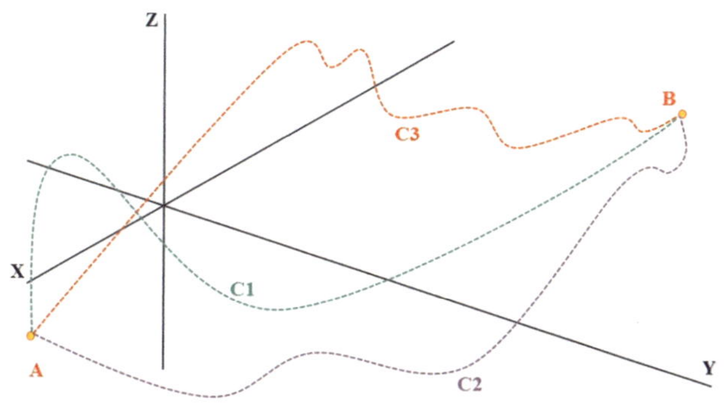

Figure 3.
Scheme of the optimal path.

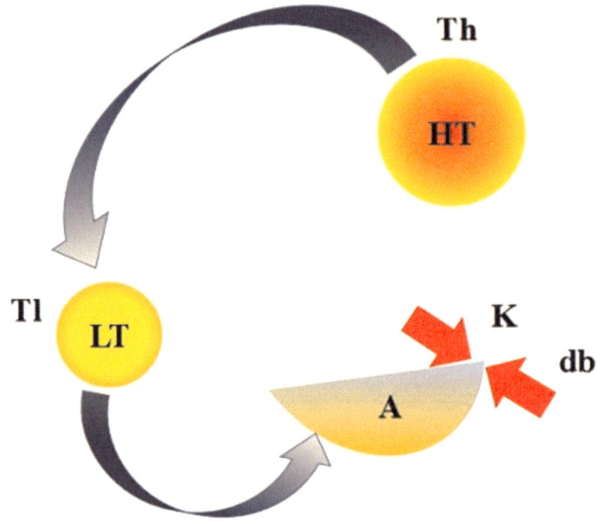

Figure 4.
Heat transfer in temperature sensor scheme.

above, moreover the mathematical analysis summarized previously helped to get the understanding of the static behavior and dynamic response of the sensor/transducer system.

In **Figure 4** is depicted a first order system based in a thermal system owing to the heating transfer has the characteristic to not have overshoots and not faster response time. Hence, Th represents the temperature of the thermal focus HT, which is bigger than Tl (temperature of the body s) whereby heat is transferred from HT to LT. Furthermore, K is the thermal resistivity, A is the section area crossed by the heating transfer and db its thickness.

Therefore, a temperatures sensor can be modeled by a first order system due to the heat transfer behavior and that model can be explained by Eq. (8) [12].

$$\Theta(t) = \Theta_f\left(1 - e^{-\frac{t}{RC}}\right) \tag{8}$$

Eq. (9) is obtained after the Laplace transformation in Eq. (8), in which K_p is the proportional gain, L is the sensor delay in temperature measurement, τ is the sensor response time for the temperature measurement [12].

$$\frac{T^{\circ}}{U(S)} = \frac{K_p e^{LS}}{\tau S + 1} \tag{9}$$

In second side, for a second order model, such as described in **Figure 5**, in which, X0 means the displacement registered by the sensor and X1 means the real displacement because of the mass M and the overshoots are depending of K and β (deformation coefficient and damping coefficient), it means that a sensor with second order response can be depicted by **Figure 5**.

Therefore, by Eq. (10) is possible to model a second order system according to understand its dynamic in time domain [12].

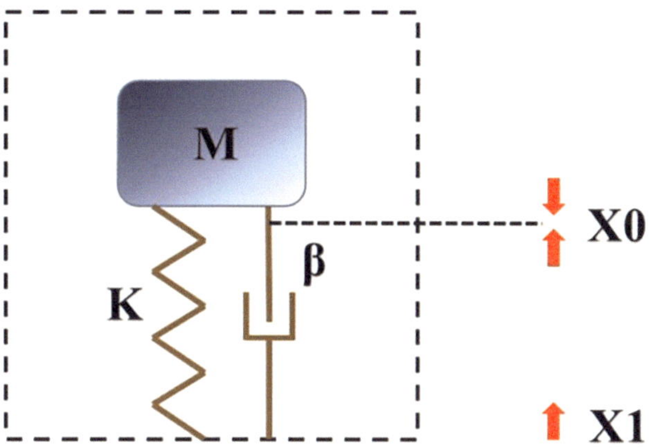

Figure 5.
Displacement in vibration sensor scheme.

$$M\left(\frac{d^2X_1}{dt^2} - \frac{d^2X_0}{dt^2}\right) = KX_0 + \beta\frac{dX_0}{dt} \tag{10}$$

By Laplace domain, it is obtained the following model due to interpret the experimental data as second order response, which is given by Eq. (11) [12].

$$MS^2X_1(S) = X_0(S)\left[K + \beta S + MS^2\right] \tag{11}$$

Thereby, Eq. (12) summarizes the parameters for a vibration sensor in Laplace domain [12].

$$\frac{X_0(S)}{S^2X_1(S)} = \frac{M}{K}\frac{\frac{K}{M}}{S^2 + \frac{\beta}{M}S + \frac{K}{M}} \tag{12}$$

The mathematical models for sensors described in paragraphs above are enhanced by adaptive models achieved from experimental analysis and the improvement of their physical parameters also can be achieved from dynamic and geometry properties by dependence of the material of the designed sensor (nanostructures).

Such as for example, the theoretical model of the temperature sensor is given by Eq. (9) that was compared by the polynomial analysis of Eq. (1) and the coefficients of the theoretical model can be compared with the MF parameters from Eq. (9), which were obtained by the measured temperature data. Moreover, for the context of the vibration sensor, its theoretical model given by Eq. (12) also was compared with the experimental information described by the parameters Eq. (9) from Eq. (1). In this context, Eq. (13) is the model of the temperature sensor, in which, its parameter "k" is the temperature sensor gain and "τ" is its response time, this expression is achieved as a consequence of the parameters correlation from Eqs. (1) and (9).

$$Gs = \frac{k}{\tau S + 1} \tag{13}$$

It is known the Zero Order Hold (ZOH) by Eq. (14).

$$G(zoh(S)) = \frac{1 - e^{-jwT_s}}{jw} \tag{14}$$

That is equivalent to Eq. (15) in Laplace domain.

$$G(zoh(S)) = \frac{1 - e^{-ST_s}}{S} \tag{15}$$

Looking for the digital model by Z transform in Eq. (16), which is achieved from Eq. (15).

$$\mathrm{H}\left(\mathrm{Z}\left\{\frac{\mathrm{k}}{\tau\mathrm{S}+1}\right\}\right) = G(Zoh)Z\left\{\frac{k}{\tau S+1}\right\} \tag{16}$$

Replacing the ZOH in last Eq. (16) is obtained Eq. (17).

$$H\left(Z\left\{\frac{k}{\tau S+1}\right\}\right) = Z\left\{\frac{1 - e^{-ST_s}}{S}\right\}Z\left\{\frac{k}{\tau S+1}\right\} \tag{17}$$

It is known $z = e^{ST_s}$ to replace in last Eq. (17) according to obtain Eq. (18).

$$H\left(Z\left\{\frac{k}{\tau S+1}\right\}\right) = (1 - Z^{-1})Z\left\{\frac{1}{S}\left(\frac{k}{\tau S+1}\right)\right\} \tag{18}$$

Eq. (19) is obtained reducing Eq. (18).

$$H\left(Z\left\{\frac{k}{\tau S+1}\right\}\right) = k(1 - Z^{-1})Z\left\{\frac{1}{S}\left(\frac{\frac{1}{\tau}}{S+\frac{1}{\tau}}\right)\right\} \tag{19}$$

By Z transform in Eq. (19) is obtained Eq. (20).

$$H\left(Z\left\{\frac{k}{\tau S+1}\right\}\right) = k(1 - Z^{-1})Z\left\{\frac{1}{S}\left(\frac{\frac{1}{\tau}}{S+\frac{1}{\tau}}\right)\right\} \tag{20}$$

Therefore, by Z transform is achieved the digital model of the proposed temperature sensor that is given by Eq. (21), because of Z transform in Eq. (20).

$$H\left(Z\left\{\frac{k}{\tau S+1}\right\}\right) = \frac{kZ^{-1}\left(1 - e^{-\frac{T_s}{\tau}}\right)}{1 - e^{-\frac{T_s}{\tau}}Z^{-1}} \tag{21}$$

By other side, the Tustin model is given by Eq. (22), in which "T_s" is the sampling time.

$$S = \frac{2}{T_s}\frac{Z-1}{Z+1} \tag{22}$$

Hence, the digital model of first order transfer function of the proposed temperature sensor (because of Tustin reduction) is given by Eq. (23) that was obtained replacing in Eq. (13) the Eq. (22).

$$H(Z) = \frac{T_s Z + T_s}{(2\tau + T_s)Z + (T_s - 2\tau)} \tag{23}$$

Eqs. (21) and (23) are the digital model for the transfer function of the proposed sensor, for which **Figure 6** shows the comparisons among the theoretical model of the temperature measurement for the operating combustion motor with the experimental temperature measurement, which were obtained and processed by the proposed temperature sensor. The blue color curve is the theoretical result based in heat transfer (Eq. (9)) from the surface motor to the sensor surface, the red color curve is the measurement data obtained by the execution of Eq. (21) in the processor of the temperature sensor, while the green color curve is the measurement data obtained by the execution of Eq. (23) in the processor of the temperature sensor. The error achieved by the measurement data represented by the red color curve was 0.5 percent approximately, and the error obtained by the measurement data represented by the green color curve was 0.9 percent approximately. For both contexts, the error analysis was made by the comparison of the measurement data with the theoretical curve (blue color).

Therefore, it can be possible to choose Eq. (23) in order to be the base of the temperature monitoring algorithm for the proposed sensor, also because it has more simple expression for the programming in comparison with Eq. (21), which has not simple elements for the programming and it can cause consequences in the computing time. However, the result is much better by the programming of Eq. (21) because it produced less error than the measurement by the processing of Eq. (23), moreover the consequence in the computing time is solved by the short response time of the sensor owing to the nanostructures characteristics of the sample that received the IR signal of the measured temperature.

In order to compare the theoretical measurement with the experimental data of the combustion motor vibration surface, there were achieved the parameters of the second order system for the proposed vibration sensor by comparison of the Eqs. (1), (7) and (12), in similar context to the temperature sensor (described in paragraphs above), it was analyzed the digital equation by Z transform and Tustin reduction according to compare with the theoretical result that is given by the blue color curve

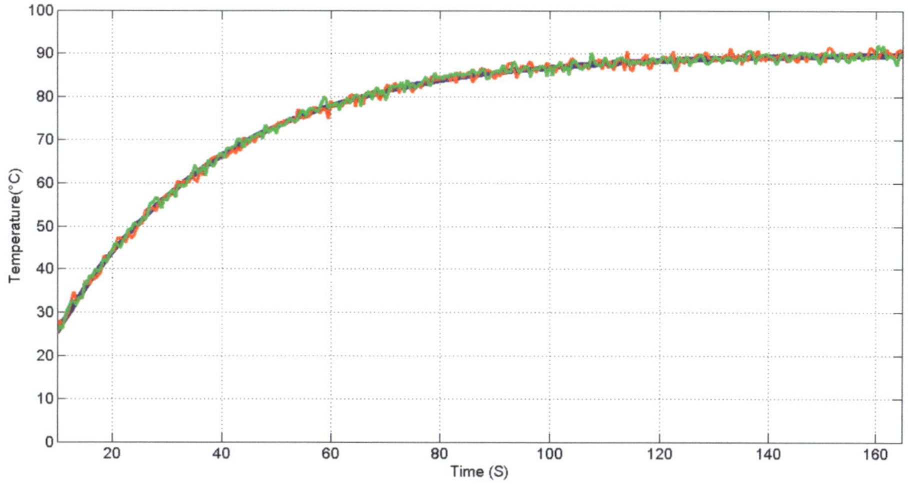

Figure 6.
Theoretical curve (blue) versus experimental curves (red and green) for the measurement temperature analysis.

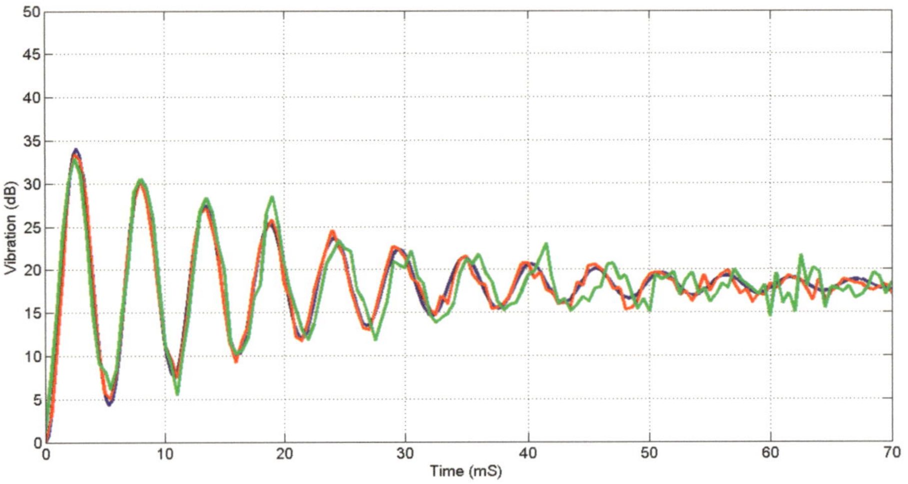

Figure 7.
Theoretical curve (blue) versus experimental curves (red and green) for the measurement vibration analysis.

and showed in **Figure 7**. The experiment was made by measuring the vibration of the combustion motor surface while it was pushed the accelerator due to keep stability of the RPM (Revolution Per Minute) and the red color curve is the data from the experimental measurement evaluated by the processor of the proposed sensor and the algorithm executed was supported by the Z transform, nevertheless, the green color curve was achieved by the experimental data that was evaluated by the Tustin model (reduction). Hence the less error value was obtained by the model based in Z transform even though the complication in its programming (in comparison of the model based by Tustin) was not a problem, because of the short response time of the sensor surface based in nanostructures.

Thus, the error for both models were less than 1 percent (Error of 0.5 percent for Z transform and 1.8 percent for Tustin reduction) and for the operating work of the combustion motor there was not necessity to use digital model expressions for the algorithms analysis of the transduction, even though the sampling time was around 200 uS (less value than the minimal response time: 2mS), hence it was continued the analysis by Laplace domain. However, it can be used for high values of operating work, maybe for future applications.

The mathematical model of sensors are evaluated also as part of a control system for vibration and temperature analysis of a combustion motor, in which was necessary to identify the system and to keep a good performance of the vibration control for a combustion motor. Therefore, an interesting evaluation is given by PID (Proportional Integral Derivative) control as part of the identification system of the combustion motor, and the physical variables (vibration and pressure) are measured by the designed sensor.

Figure 8 shows the adaptive cascade algorithm that is represented by a block diagram to suit the measured signal received from the vibration motor. The input signal is given by the IR measured signal U1 that is adjusted by the matrix weights W1 and M1, controlled by C1 over the sensor/transducer P1.

The response signal U2 (electrical value of the IR signal measured from the surface motor) is controlled by C2 and adjusted by M2, W2 due to obtain the temperature transduced as a consequence to know the function P2, moreover U3 (which also is U2)

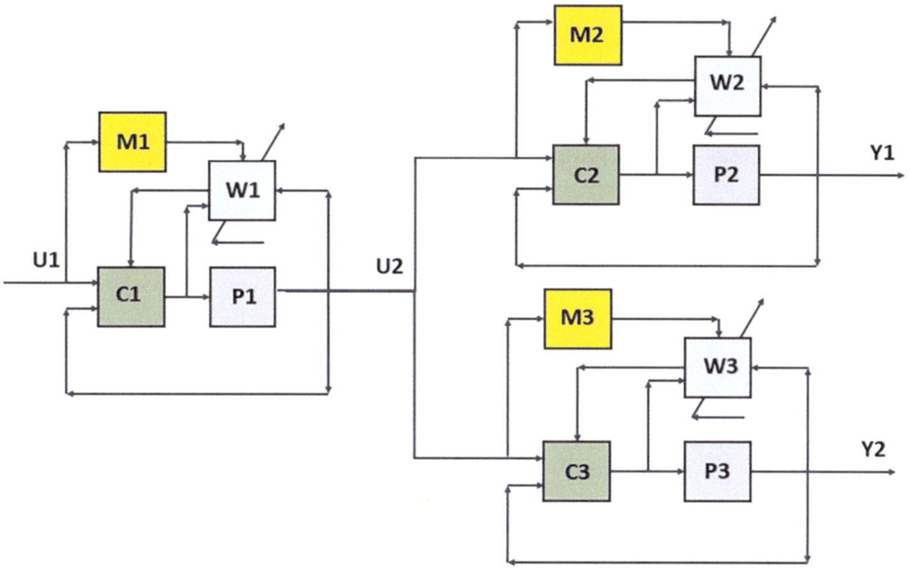

Figure 8.
Block diagram scheme for the main control algorithm.

in concurrent time is compensated by C3 and adapted by M3, W3 in order to achieve the vibration response of the designed sensor as a consequence to know the transfer function P3.

The internal controller PID in block diagram scheme is depicted by **Figure 9**, which as necessary for the identification system of the combustion motor parameters that are part of the motor adaptive control. The input signal In(S) gives information of the desired value in temperature and vibration of the motor surface, Con(S) is the PID control (parameters), Pt(S) is the transfer function of the plant (motor surface), S(S) is the transfer function of the designed sensor, and Y(S) is the response signal.

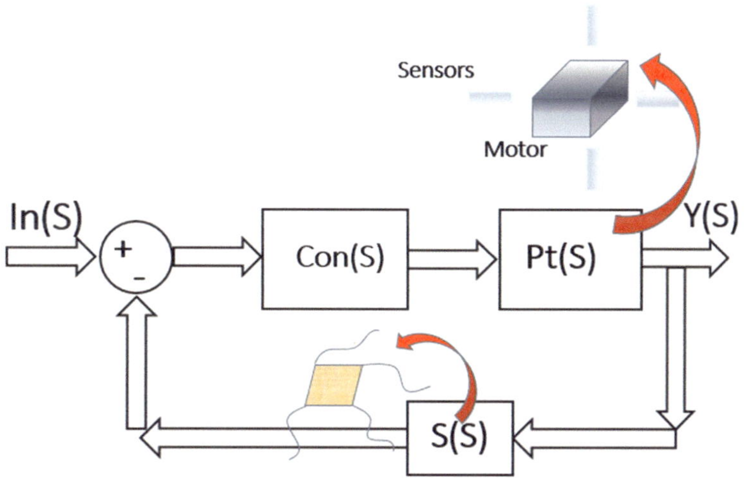

Figure 9.
PID controller used for the identification system.

Eq. (24) is obtained as a result of the algebra analysis from the block diagram above.

$$(In(S) - S(S)Y(S))Con(S)Pt(S) = Y(S) \tag{24}$$

Thus, the transfer function obtained from the reduction of Eq. (24) is given by Eq. (25).

$$\frac{Y(S)}{In(S)} = \frac{Con(S)Pt(S)}{1 + S(S)Con(S)Pt(S)} \tag{25}$$

In Eq. (26), it is generalized the transfer function for temperature/vibration combustion internal control, in which K_S is the gain parameter of the designed sensor, τ_s is the response time of the sensor. K_P is the controller proportional gain, K_D is the Controller derivative gain, K_I is the Controller integral gain. K_{pt} is the gain parameter of the plant (combustion motor), τ_{pt} is the plant response time.

$$\left(\frac{K_S}{\tau_s S + 1}\right)\left(K_P + K_D S + \frac{K_I}{S}\right)\left(\frac{K_{pt}}{\tau_{pt} S + 1}\right) + 1 = 0 \tag{26}$$

The reduction from the equation is given by Eq. (27), for which was decided a Proportional Derivative (PD) controller due to achieve a fast control response under the vibration motor.

$$\tau_s \tau_{pt} S^2 + \left(\tau_s + \tau_{pt} + K_s K_D K_{pt}\right)S + \left(K_s K_p K_{pt} + 1\right) = 0 \tag{27}$$

In Eq. (28) is organized Eq. (27) as a polynomial in second descending order.

$$S^2 + \frac{\left(\tau_s + \tau_{pt} + K_s K_D K_{pt}\right)}{\tau_s \tau_{pt}} S + \frac{\left(K_s K_p K_{pt} + 1\right)}{\tau_s \tau_{pt}} = 0 \tag{28}$$

The control parameters can be obtained by different methodologies such as the stability analysis, furthermore the comparison with the theoretical model of the system dynamic given by Eq. (29) [3, 12], in which ω_0 is the natural frequency for the system and ϵ is the damping effect.

$$S^2 + (2\epsilon\omega_0)S + {\omega_0}^2 = 0 \tag{29}$$

Hence, the control parameters K_p and K_I, can be obtained by the comparison of the coefficients from Eqs. (28) and (29), from which are proposed the following Eqs. (30) and (31) that are functions of τ_s,τ_{pt},K_s,K_{pt},ω_0 and ϵ.

$$\frac{\left(\tau_s + \tau_{pt} + K_s K_D K_{pt}\right)}{\tau_s \tau_{pt}} = 2\epsilon\omega_0 \tag{30}$$

$$\frac{\left(K_s K_p K_{pt} + 1\right)}{\tau_s \tau_{pt}} = {\omega_0}^2 \tag{31}$$

Also, Eq. (32) is the proportional parameter of the PD controller obtained from Eq. (31).

$$K_P = \frac{\omega_0{}^2 \tau_s \tau_{pt} - 1}{K_s K_{pt}} \tag{32}$$

Finally, the derivative gain is obtained from Eq. (30) and showed by Eq. (33).

$$K_D = \frac{2\varepsilon\omega_0 \tau_s \tau_{pt} - \tau_s - \tau_{pt}}{K_s K_{pt}} \tag{33}$$

After to obtain the control parameters, it is possible to warrant the influence of the designed sensor in the stability of the system, thus, analyzing Lyapunov stability from equation previous, for which Eq. (34) is the complement of Eq. (17) in which U(S) is the input excitation signal and R(S) is the small displacement in Laplace domain.

$$S^2 + \frac{(\tau_s + \tau_{pt} + K_s K_D K_{pt})}{\tau_s \tau_{pt}} S + \frac{(K_s K_p K_{pt} + 1)}{\tau_s \tau_{pt}} = \frac{U(S)}{R(S)} \tag{34}$$

Eq. (35) is a reduction from Eq. (34) but in time domain.

$$\frac{d^2 r(t)}{dt^2} + \frac{(\tau_s + \tau_{pt} + K_s K_D K_{pt})}{\tau_s \tau_{pt}} \frac{\mathrm{d}r(t)}{dt} + \frac{(K_s K_p K_{pt} + 1)}{\tau_s \tau_{pt}} r(t) = u(t) \tag{35}$$

In addition, preparing variable changes and showed by Eq. (36).

$$\mathrm{y(t)} = \frac{\mathrm{d}r(t)}{dt} \tag{36}$$

Eq. (37) is achieved replacing the Eqs. (36) and (34) in Eq. (35), for u(t) null:

$$\frac{\mathrm{d}y(t)}{dt} = -\frac{(\tau_s + \tau_{pt} + K_s K_D K_{pt})}{\tau_s \tau_{pt}} y(t) - \frac{(K_s K_p K_{pt} + 1)}{\tau_s \tau_{pt}} r(t) = 0 \tag{37}$$

Organizing the last equation by energy analysis, $\boldsymbol{E(r, y)}$, in order to find the Lyapunov equation, which is positive and can achieve the first Lyapunov condition given by Eq. (38).

$$E(r, y) = 0.5(y(t))^2 + 0.5 \frac{(K_s K_p K_{pt} + 1)}{\tau_s \tau_{pt}} (r(t))^2 = 0 \tag{38}$$

Looking for the second Lyapunov condition by the inequality (39).

$$\frac{\mathrm{d}E(r(t), y(t))}{dt} \leq 0 \tag{39}$$

Therefore, the inequality (40) is obtained replacing Eq. (38) in the inequality (39).

$$y(t) \frac{\mathrm{d}y(t)}{dt} + \frac{(K_s K_p K_{pt} + 1)}{\tau_s \tau_{pt}} r(t) \frac{\mathrm{d}r(t)}{dt} \leq 0 \tag{40}$$

Also, replacing Eqs. (36) and (37) in the inequality (40) is obtained the inequality (41).

$$y(t)\left(-\frac{(\tau_s+\tau_{pt}+K_sK_DK_{pt})}{\tau_s\tau_{pt}}y(t)-\frac{(K_sK_pK_{pt}+1)}{\tau_s\tau_{pt}}r(t)\right)+\frac{(K_sK_pK_{pt}+1)}{\tau_s\tau_{pt}}r(t)y(t)\leq 0 \tag{41}$$

Finally, it is obtained the inequality (42) due to achieve the second Lyapunov condition, moreover while $\boldsymbol{\tau_s}$ is small the control system get better stability, it can be possible by sensors with short response time such as the sensors based in nanostructure (as it is designed the proposed sensor of this research).

$$-\frac{(\tau_s+\tau_{pt}+K_sK_DK_{pt})}{\tau_s\tau_{pt}}y(t)^2\leq 0 \tag{42}$$

All the analysis was made in Laplace domain, because the response time is enough bigger than the sampling time of both systems "temperature and vibration" by the processor of the advanced sensor. Furthermore, the robustness and short response time of the sensors based in nanostructures give possibility to execute complicated algorithms, however this computing task can be prioritized for future analysis.

4. Sensor design

A good sensitivity of the temperature and vibration of the combustion motor surface depends of the internal membranes of the designed smart sensor, which are prepared by nanotructures of AAO, hence the elaboration of the nanostructures (nanoholes and nanotubes) are prepared by a sequence of steps over aluminum: ultrasound cleaning of aluminum, electropolishing, anodization, and atomic load deposition. Nevertheless, if there is not a good quality of aluminum, the base of the nanostructure could not have robustness over their geometry, in **Figure 10** is showed some aluminum samples at 99.9 percent for the designed vibration sensor [13]. Sensors based in samples of nanostructures give the possibility to achieve robustness and short response time [14].

Figure 10.
Aluminum samples.

Figure 11.
Electropolishing.

Electropolishing is an electrochemical process according to achieve more sophisticated cleaning over every sample by electrolysis and from anode to cathode [13], that is showed by **Figure 11**.

After the electropolishing, the samples achieve very much shining like a mirror whereby **Figure 12** shows 4 samples, in which one of them (up left side) is showed the best electropolished sample.

The anodization produces chemical effects on the cleaned aluminum such as the holes in nanoscales, this process was made in controlled electro-chemical perchloric acid reaction, and adjusting the electrical source between 20 V to 30 V in environment around 0 Celsius degrees, which is showed by **Figure 13**. Moreover, by electrochemical deposition was possible to prepare nanostructures amorphous over the AAO samples.

As a result were obtained nanostructures samples based in AAO, which are showed by **Figure 14**. The subfigures "A" and "C" are the samples prepared in the Applied Nanophysics, Institute for Physics of TU Ilmenau by the cooperation research between PUCP and TU Ilmenau, "B" is the sample prepared in the researching

Figure 12.
Electropolished samples.

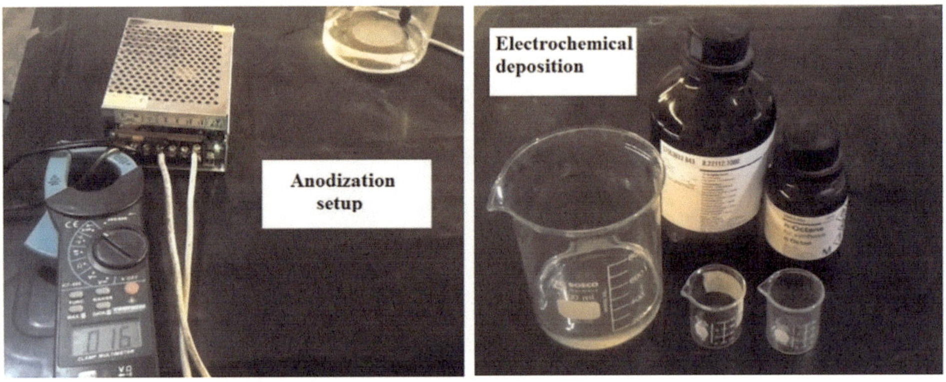

Figure 13.
Anodization and electrochemical deposition process for the sensor elaboration.

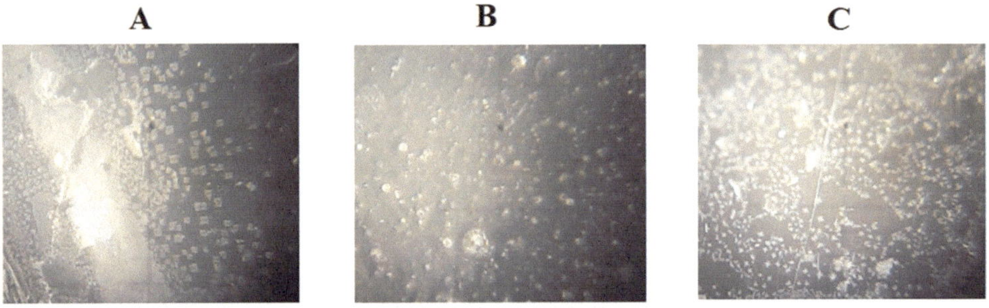

Figure 14.
AAO samples for the sensor design.

laboratories 1 and 2 of the Mechanical Department of PUCP by the optimal procedure discovered by Prof. Lei group.

The samples prepared are adapted through its own geometry in nanoscale due to obtain amorphous nanotubes for the vibration sensor and amorphous nanoholes for the temperature sensor. There is an IR emitter as part of the sensor/transducer design, which send IR signal to the combustion motor surface in controlled frequency due to recognize the differences with the IR signal caused by the temperature changes of the combustion motor surface. Therefore, the temperature measurement is correlated between the IR signal with its own IR signal caused because of the temperature changes in the motor surface (temperature measured by IR [6]).

The receptors are given by the nanostructures samples that send the measured data (temperature and vibration) to the microcontroller owing to execute the mathematical model of the correlation among the theoretical model with the experimental mathematical analysis, finally that data is sent to the user by wireless port. In order to obtain the electrical response of the measured variable, it was fixed some cables in every corner of the sample. The electrical resistance equivalent is the physical variable to correlate the measured variable. In **Figure 15** are showed 3 prototypes of designed sensors/transducers (A, B and C).

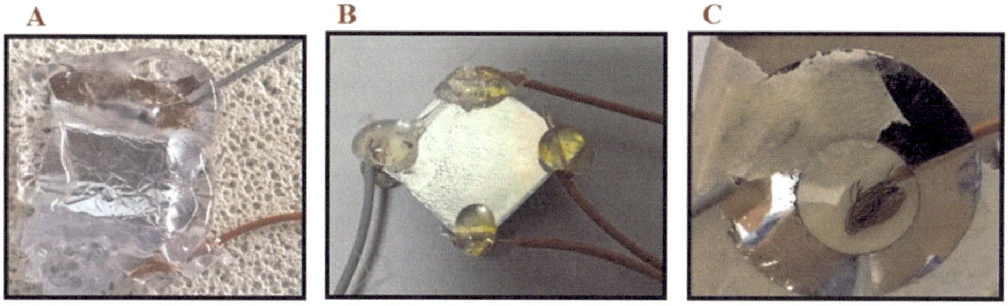

Figure 15.
Prototypes of proposed sensors/transducers.

The sample B of **Figure 12** is showed under a microscope Litz in the scale 25 micrometers, thereby the **Figure 16** shows some amorphous structures with maximal scales are around 1000 nanometers.

In **Figure 17** is depicted the algorithm scheme for the sensor transduction by the flowchart of the sensor/transducer operation described in paragraphs above, in which physical variables temperature and vibration are measured and processed by concurrence and finally both signals are correlated according to obtain the final transduction result.

Algorithm scheme in operation during the measurement of the physical variables vibration and temperature of motor surfaces is depicted in the following **Figure 18**, thereby the predictions of the adaptive algorithm can be obtained by the interpretation of the IR reception [6].

In the following **Figure 19**, by the curve A is depicted the measurement data from the designed sensor, the curve B represents the measurement data received by the personal computer at 50 meters of distance with a delay L1 because of the medium used was internet. Nevertheless, if it is used radiofrequency medium communication the delay is reduced in L2 as it is depicted by the curve C.

It is showed by the **Figure 20** the motor used for the experiments, which is a Nissan frontier 2005. The sensors (4 of them) were positioned around the motor by non-contact in every Cartesian axis (X, Y, Z).

Figure 16.
Amorphous nanostructures over the surface of the designed sensor/transducer.

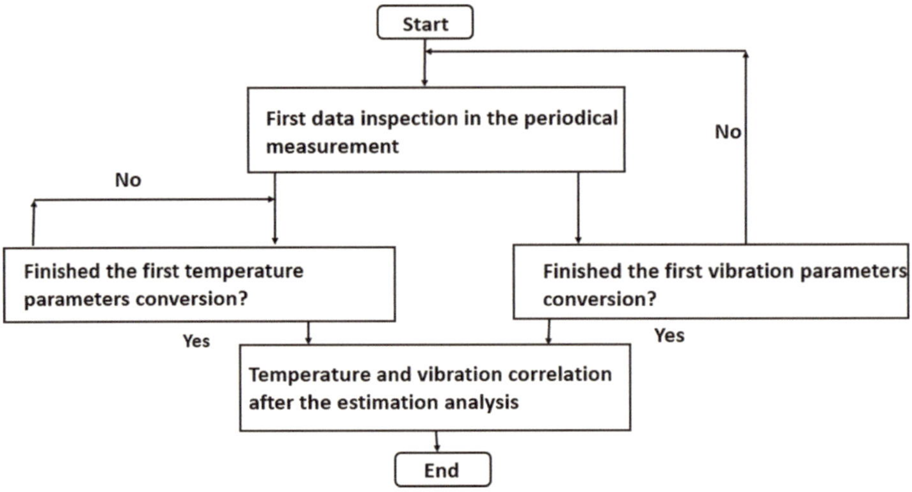

Figure 17.
Algorithm scheme for the sensor transduction.

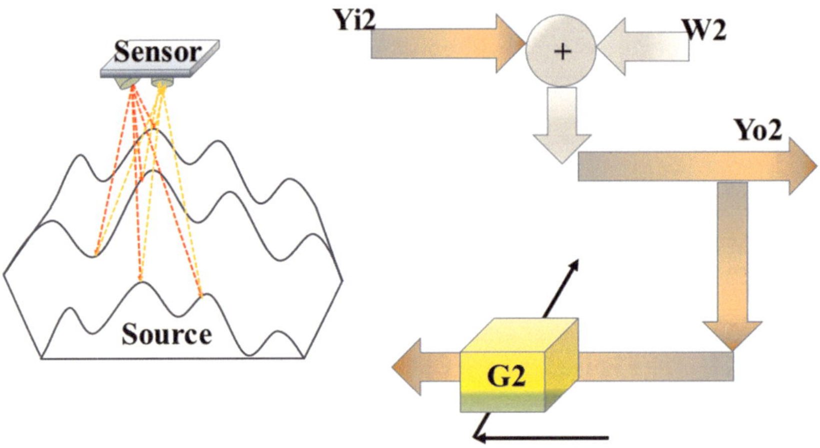

Figure 18.
Algorithm scheme in operation during the measurement.

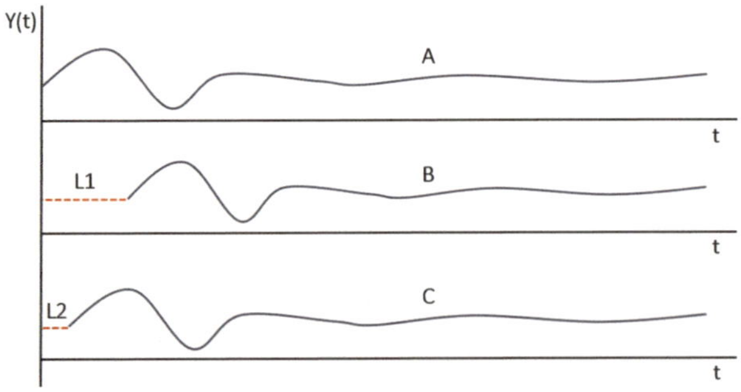

Figure 19.
Representation of the measured data (a), its transmission by internet (B) and by radiofrequency (C).

Figure 20.
Combustion motor of a Nissan frontier 2005, in which were made the experiments.

Fixing a small electro-pneumatic actuator over the accelerator pedal, which receive the control signal according the main control algorithm that receive the vibration and temperature signal from the designed sensor (through the radiofrequency antennas).

In **Figure 21** is showed the vibration of the combustion motor measured by the designed sensor. The vibration signal was transduced from IR to electrical signal (Voltage) that is showed by the blue color curve, and its amplification in equivalent of Decibels by the red color curve in the **Figure 18**. The combustion motor was evaluated in operation around 4000 RPM and the maximal spectral density was obtained approximately in 68 Hz that can be seen by the green color curve, which in addition can justify the operation frequency of the internal combustion motor.

The vibration of the motor surface was captured by the designed sensor/transducer and it was sent by IR to the emitter antenna that sent the data by radiofrequency to the receptor antenna, which is at 50 meters outside, moreover the receptor antenna sent

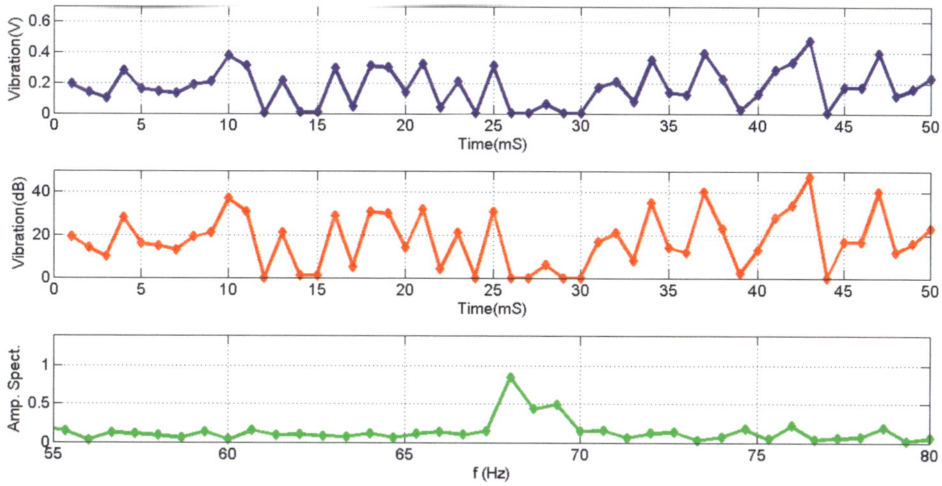

Figure 21.
Vibration curves achieved by the designed sensor.

the measured vibration to a personal computer by IR, also, according the control signal to activate the electro-pneumatic actuator to change the position of the accelerator pedal. That curves information can be interpreted by the user according to get understanding of the behavior of the motor as a consequence of the combustion. It is necessary to remind that the intelligent algorithm of the adaptive correlation to achieve the physical variables transductions had 87 SNR (Signal to Noise Ratio) in average.

The operating frequency is quite dependent of the vibration frequency of the combustion motor, because of this is the main source of changes in the described system. The vibration measurement can be simpler due to its correlation with the vibrating motor, however, the temperature measurement depends not only from the operating vibration motor, it also depends from its delay caused due to the thermal inertia. Furthermore, the sample frequency is part of the designed sensor analysis.

It was evaluated the performance of the sensor/transducer by different temperature changes, which were caused by accelerating the motor during 45 minutes approximately. **Figure 22** shows the optimal estimated temperature of the surface motor that is given by the green color curve. The optimal estimation is achieved as a consequence of the correlation between the experimental measurement (blue color curve) with the temperature measurement by a thermocouple type k (red color curve). Hence, the designed sensor/transducer can measure the temperature of the motor surface by optimal estimations according to answer in front of disturbances, moreover the measured data was sent through IR to the emitter antenna that sent the data to the receptor antenna by radiofrequency at 50 meters outside, from which the information is received by a personal computer through IR with the receptor antenna. The user can interpret the data received, such as for example the diagnostic of the motor by the combustion effects because of the temperature changes. The delay obtained by the radiofrequency data monitoring (vibration and temperature of the surface motor) was between 700mS to 800mS, and the delay obtained by internet data monitoring was between 1.2S to 1.4S.

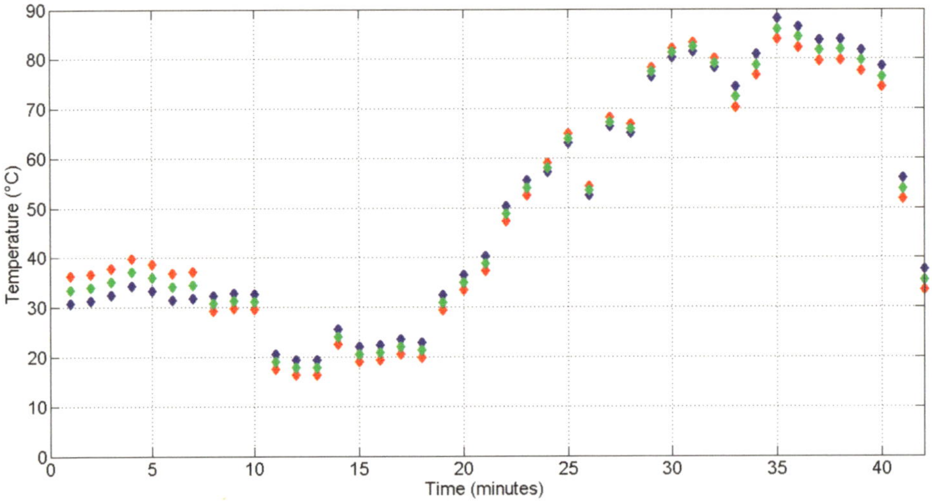

Figure 22.
Temperature curve from the designed sensor in Celsius degrees.

5. Conclusions

It was designed an intelligent temperature and vibration sensor/transducer based in nanostructures of AAO owing to measure the temperature and vibration over the surface of combustion motors, the achieved data is sent to the user according to get a diagnostic of the combustion motor performance and the user can understand whether the motor could need reparation or not.

The novelty of this proposed article is given in the mathematical analysis to design sophisticated transducers, the support of the analysis is the polynomial structure of the mathematical modeling due to the correlation between the theoretical equation of heating and vibration transfer with the experimental data achieved during the calibration. The MF in adaptive coefficients of the final model gives the advantage to optimize the data filtering while it is supported by the calibration information.

The algorithm designed as a consequence of the mathematical model can be programmed by different language programming because of the simple instructions and the weights achieved from the mathematical model and calibration data help to adapt the measured data according to estimate the right measured temperature and vibration by non-contact transduction. Hence, the transducer designed optimize electronic components of instrumentation, and as a consequence can enhance the effect over pollution caused by combustion motors in Peru, which are used in big quantities by the public and private transport without a continuous and practical monitoring of their operation, moreover it was possible to evaluate the performance of the designed sensor through wireless communication by IR and radiofrequency, which was possible to achieve because of the short response time and robustness of the designed sensor give enough time for the data communication according to get telemetric monitoring of the measured variables.

Finally, the designed sensor can use the energy stored from its own sun energy converter, which gives more independence and autonomy to the designed sensor.

The sample time was also part of the analysis because of it was obtained the digital models for the equations that can be programmed by the processor of the intelligent sensor (to measure the temperature and vibration of the motor surface). Nevertheless, for the RPM operating work the achieved error was less than 1 percent (for the Z transform reduction), hence it was decided to prioritize the equation analysis by Laplace domain, but the model can be used in their equivalent digital expressions whether the total computing response time could be near the system response time (temperature and vibration of the computer motor) even though the fast response in the transductions because of the sensor is based in nanostructures is a good advantage for the communication time when it was sent the measurement by wireless.

6. Future work

It is proposed for a future research that the designed transducer can improve its performance for more amplitude of range of work due to the mathematical analysis can adapt the characteristics parameters of the sensor during the calibration, furthermore, the adaptive transduction can give faster and good response in nonlinear range of work. In addition, with the mathematical model designed for the vibration/temperature transducer can be adapted for complex correlations or control tasks that could be made on wireless.

Moreover, it is proposed for a future research to enhance the applications of the designed sensor in telemetry control and increase the distance between the antennas which support for the measured data transmission.

The heat produced due to the combustion motor operation (the maximal temperature value of the motor surface) can be used to be transformed in electrical energy to be used by the control system, thus this is a target for a future work of this research.

Acknowledgements

It is expressed deep warm gratefulness to Mrs. Aleksandra Ulianova de Calderón due to her total support for the development of this research due to find the understanding of the compromise among new technologies with the environment cares.

There is expressed special thankful to the Mechatronic Engineering Master Degree Program at PUCP, to the Engineering Department PUCP, and DGI ("Dirección de Gestión de la Investigación") researching office from PUCP because of its financial support in this research through the financing FONCAI.

Author details

Jesús Alan Calderón Chavarri[1,2]*, Julio César Tafur Sotelo[2], Eliseo Benjamín Barriga Gamarra[2], John Hugo Lozano Jáuregui[2,3], Dante Jim Randal Gallo Torres[2], Rodrigo Alonso Urbizagástegui Tena[2], Jaime Eduardo Zeña Delgado[4] and Christian Enrique Gózar Pastor[2]

1 Applied Nanophysics, Institute for Physics, Technical University of Ilmenau, Ilmenau, Germany

2 Control Engineering and Automation Master Program, Mechatronic Engineering Master Program, Engineering Department, Pontifical Catholic University of Peru, Peru

3 Northen (Artic) Federal University named after MV Lomonosov, Arkhangelsk, Russian Federation

4 National Engineering University (UNI), Industrial Engineering Department, Peru

*Address all correspondence to: alan.calderon@pucp.edu.pe

References

[1] Landau LD, Lifshitz EM. Theory of Elasticity, Course of Theoretical Physics. Vol. 7. Institute of physical problems, Great Britain: USSR academy of science; 1959

[2] Richard F, Robert L, Matthew S. The Feynman lectures on physics. Vol. I. The United State of America: New millennium editors; 1962

[3] Person AE. Aerodynamic parameter estimation via Fourier modulating function techniques. USA: Center for Aero Space Information; 1995

[4] Zhang F, Jiang M, Zhang L, Ji S, Sui Q, Chenhui S, et al. Internal combustion engine fault identification based on FBG vibration sensor and support vector machines algorithm. Mathematical Problems in Engineering. Vol. 2019, 2019. DOI: 10.1155/2019/8469868

[5] Christiana B, Volkb J, Lukàcsb IE, Sautieffc E, Sturmd C, Graillote A, et al. Piezo-force and vibration analysis of ZnO nanowire arrays for sensor application. Procedia Engineering, 2016; **168**:1192-1195

[6] Optris, Infrared Thermometers. Basic Principles of Non-Contact Temperature Measurement. Germany: GmbH; 2022. Available from: https://www.optris.de/infrarot-thermometer

[7] T. E. Sensor Solutions. TS318-11C55 Thermopile Sensor, Germany. 2015. Available from: https://www.te.com/usa-en/product-G-TPCO-032.html

[8] Designing a Low Cost, High-Accuracy Infrared Thermometer, application note. Texas. USA: Texas Instruments; 2020

[9] NDT, Tek Know. Innovative technologies. Non destructive devices control. Russian Federation. 2015. Available from: https://www.tek-know.ru/catalog/po-vidam-izmereniy/temperatura/

[10] He J-H, Liu D-P, Chung C-H, Huang H-H. Infrared thermography measurement for vibration-based structural health monitoring in low visibility harsh environments. Sensors MDPI. 2020;**20**

[11] Lei Y, Cai W, Wilde G. Highly ordered nanostructures with tunable size, shape and proerties: A new way to surface nano-patternin using ultrathin alumina masks. Progress in Materials Science. 2007;**52**:465-539. DOI: 10.1016/j.pmatsci.2006.07.002

[12] Calderón Ch. J. Alan, Barriga B, Tafur J, Lozano J, Lengua JC, Solano G, et al. Optimal vibration analysis for a combustion motor. In: IEEE Conference on Industrial Electronics and Applications (ICIEA). IEEE; 2021

[13] Lei Y, Cai W, Wilde G. Highly ordered nanostructures with tunable size, shape and properties: A new way to surface nano-patterning using ultra-thin alumina masks. Progress in Materials Science. 2007;**52**:465-539

[14] А. А. Теплоухов, Н. А. Семенюк, Д. А. ОСНОВЫ СИНТЕЗА НАНОСИСТЕМ Федеральное государственное бюджетное образовательное учреждение высшего образования «Омский государственный технический университет». 2020

Chapter 3

Perspective Chapter: Predicting Vehicle-Track Interaction with Recurrence Plots

Juan Carlos Jauregui-Correa

Abstract

This chapter presents a method for identifying the dynamic response of vehicles (railcars) running on a rail track. The method is based on the recurrence plots developed from the phase diagram (phase plane). The phase plane is constructed using Hamilton's principle for a single-mass system. The state variables are estimated from acceleration measurements registered in an experimental test rig. The measurements were recorded with accelerometers and gyroscopes mounted on the railcar that circulated on a closed-loop track. The acceleration data were integrated using the empirical mode decomposition method and the time delay principle. The acceleration data were separated into two data vectors: rigid body motion and vibration modes. The recurrence plots were built for both data vectors, and it was found that the vibration modes were more sensitive to track defects and curvature changes. The recurrence plots were analyzed with the recurrence quantification analysis, and it was found that the Recurrence Plots can determine the location and type of defects.

Keywords: recurrence plots, recurrence quantification analysis, vehicle dynamic measurements, empirical mode decomposition, phase plane

1. Introduction

Many researchers work on the life prediction of nonlinear systems. Although the topic has been under research for decades, there are still many uncertainties and doubts, and still, it is an open issue from a practical point of view. There are different alternatives for modeling and analyzing nonlinear systems. This chapter presents the application of recurrence plots to predict defects in rails and railcars.

Recurrence plots are based on Poincaré's concepts. Eckmann et al. [1] worked further on Poincaré's principles and defined the basic procedure for constructing recurrence plots from a phase plane (phase space). Marwan and Weber [2] and Webber et al. [3] represented different dynamic systems using the recurrence plot procedure; their primary contribution is that the trajectory along a phase plane could be quasi-stationary.

The work presented by Eckmann et al. [1] described the application of recurrence plots to determine the time constancy of dynamic systems. They were the first to distinguish that the recurrence plots can measure the entropy of the phase plane, the dimension spectrum, or other information dimensions. They constructed phase plane orbits and estimated the repeatability of each cycle; then, they quantified the number of times that a point appeared in different cycles and proposed a method for finding time correlations in a signal. They distinguished two characteristics in the recurrence plots: large-scale forms "topologies," and small-scale forms "textures." They illustrated their results with experimental and numerical data. A detailed description of "topologies" and "textures" are presented in the following sections.

Many publications deal with the application of recurrence plots to single-frequency signals. For this kind of signal, the phase plane can be constructed by using the shifting process. But, in most cases, the dynamic response combines different frequencies, transient responses, and nonlinear effects. Torres et al. [4] analyzed the error caused by applying the shifting process to nonlinear signals. To avoid this error, Torres et al. [5] proposed a different alternative. This procedure is further discussed in the following sections.

Recurrence plots have been applied to electroencephalogram signals (EEG) [6–9]. They also have been applied to direct current discharge plasma and the identification of the geodesic distance on Gaussian manifolds for chaotic systems [10]. Kwuimy and Kadji [11] and Kwuimy et al. [12–14] applied the recurrence plots to two Van der Pol type oscillators coupled by a nonlinear spring. They estimated the synchronization using two Recurrence Plots. Similar results are obtained with the Kuramoto's parameter. Jana et al. [15] represented a food chain system as nonlinear ordinary differential equations, and they applied the Recurrence Plot for identifying the dynamic parameters.

Several researchers have analyzed data generated with a Rössler system. Thiel et al. [16] embedded Gaussian noise into the Rössler model and identified the noise with the recurrence plot. Kiss et al. [17] identified synchronization on a set of Rössler oscillators using recurrence plots and determined the synchronization by calculating the cross-correlation and the probability of two-state positions coinciding in the same phase plane after a certain period. Prakash and Roy [18] represented a chaotic electric system with a Rössler model.

Recurrence Plots have been used to explain the dynamic characteristics of bubbles within a water flow [19] and to flow measurements [20]. Xiong et al. [21] compared the empirical mode decomposition and the Recurrence Plots for the analysis of traffic flow, and Tang et al. [22] proposed an intelligent traffic control system based on the Recurrence Plots. Vlahogianni and Karlaftis [23] determined the complexity of traffic flow time series using the Recurrence Plots. Ukherjee et al. [24] identified the dynamic behavior of wireless network traffic by utilizing the Quantification analysis of Recurrence Plots. Syta and Litak [25] identified cutting parameters in a machining process with Recurrence Plots. In a similar work, Elias and Namboothiri [26] identified chatter in a turning process for constructing the phase plane and found the time delay by using the average mutual information function.

The recurrence plots show patterns with specific topologies and textures that reflect the system's dynamic behavior. The analysis of these patterns relates the topologies and textures to the system's response. In this context, Leonardi [27] measured the entropy of the Recurrence Plot to signals without noise. Spiegel et al. [28] described different measures and analyses for Recurrence Plots. Belaire and Contreras [29] classified signal data into a set of embedded vectors separated by a time delay. This concept is only valid for time series that have a single frequency. Pham and Yan [30] proposed the sample entropy to measure Recurrence Plots irregularities. Girault [31] measured the symmetry in Recurrence Plots to identify the system's dynamic behavior. Sipers et al. [32] defined a procedure for retrieving a signal from a

Recurrence Plot, but their method is valid only for embedded signals. Bot et al. [33] used Recurrence Plots to detect unknown signals embedded in white noise.

The life prediction of rails and railcars depends on the ability to determine failures at the rail, the wheel, and the rotating elements of the railcar. The difference from other systems is that these variables are a function of time and location, and the wheel-rail interface determines the dynamic condition. The rail's imperfections and train speed are the dominant factors affecting the dynamic load at the wheel-rail contact point. Therefore, it is necessary to identify the dynamic load and its location along the track.

Ngamkhanong et al. [34] reviewed different models for describing the wheel-track interaction and the complexity of the elastic interaction between the rotating elements and the substructure. Most models assume that the rail behaves as an elastic beam supported by individual springs (sleepers) [35]. Ciotlaus et al. [36] defined the interaction based on the rolling contact, fatigue, and wear. These models confirmed the fatigue failures described by Smith [37].

This chapter presents the application of the Recurrence Plots to identify the dynamic effect of a rail on a railcar. The data were obtained in an experimental rig, and they consisted of acceleration and velocity measurements registered at the railcar. The acceleration was integrated using the empirical mode decomposition method and the shifting property of periodic functions [5]. Results can be extrapolated to real measurements.

The Recurrence Plots were produced from acceleration data obtained with a triaxle MEMs sensor and a triaxle gyroscope. It was possible to reproduce the railcar movements in every direction (6 degrees of freedom) with these data. Since the railcar travels in a close circuit, the acceleration data recorded rigid body and vibration motions. The data were regrouped into nonperiodic motion (rigid body motion) and periodic motion (vibrations) to separate the two types of motions. The new data sets were processed with the empirical mode decomposition method.

2. Recurrence plot

The basis for constructing recurrence plots is the phase diagram or phase plane. The phase plane [38] describes the stability of a dynamic system. It determines the relationship between the potential and kinetic energies at a given time interval and predicts the evolution of the energy balance. The recurrence plot is constructed from the analysis of the evolution of a phase plane. The dynamic system evolves following different paths represented in phase planes; these planes are built at intervals defined by the fundamental period. Then the recurrence plot identifies the number of times that a phase plane vector (the state variables that describe the system dynamics) repeats or recurs at every fundamental period. Before describing the construction of a recurrence plot, the following section introduces the basis of the phase plane.

2.1 Phase plane

A phase plane describes the mass trajectory along a two-dimensional energy field. (Two-dimensional state variables). Representing the dynamic behavior of a linear system as a differential equation system:

$$\dot{\boldsymbol{x}} = \boldsymbol{B}\boldsymbol{x} \tag{1}$$

Matrix ***B*** contains the system's parameters, and it is derived from Hamilton's principle:

$$H(p,q) = T(p) + V(q) \tag{2}$$

Or

$$T(p) = \frac{p^2}{2m} \tag{3}$$

Hamilton's principle states that the equilibrium of the system is obtained when:

$$\dot{q} = \frac{\partial H}{\partial p} \tag{4}$$

And

$$\dot{p} = -\frac{\partial H}{\partial q} \tag{5}$$

Therefore, at any instant, there is a function $\phi(p,q)$ such that

$$\frac{d\phi}{dt} = \frac{\partial \phi}{\partial q}\frac{\partial H}{\partial p} - \frac{\partial \phi}{\partial p}\frac{\partial H}{\partial q} \tag{6}$$

The dynamic stability and the evolution of the phase plane can be derived from Eq. (6). If, in the phase plane, the function ϕ is constant in any trajectory, then the system is stable (Liouville's theorem).

$$\frac{dH}{dt} = \frac{\partial H}{\partial q}\dot{q} + \frac{\partial H}{\partial p}\dot{p} = 0 \tag{7}$$

Eq. (7) implies that the phase plane has a constant volume.

If the mass is constant, the linear momentum only depends on the mass velocity. In linear systems, the potential energy is proportional to the displacement; therefore, the phase plane can be represented with the state variables velocity and position. **Figure 1** shows the phase plane of a single degree of freedom system with a harmonic response ($\dot{p} + kq = 0 \; and \; \dot{q} = \frac{p}{m}$). It is clear that the trajectory is constant and smooth; meanwhile, a nonlinear system will show irregular patterns. These irregularities are the basis for studying nonlinear systems with the recurrence plots.

2.2 Definition of the recurrence plot

The phase plane function can be defined as a vector that depends on a parameter (time) and defines the state of the dynamic system (**Figure 1**):

$$\overline{x}(t) = [\overline{x}_1, \overline{x}_2, \ldots, \overline{x}_n] \tag{8}$$

The dynamic evolution in time is represented in **Figure 2**. If the dynamic response is steady, then $\overline{x}_1(t) = \overline{x}_n(t + n\tau)$, where τ is the period of a harmonic system and n is an integer.

A system recurs if two subsequent states are equal and it repeats every period τ. According to Eckmann, Kamphorst, and Ruelle [1], a recurrence plot is a matrix representation of the similarity of two consecutive state conditions:

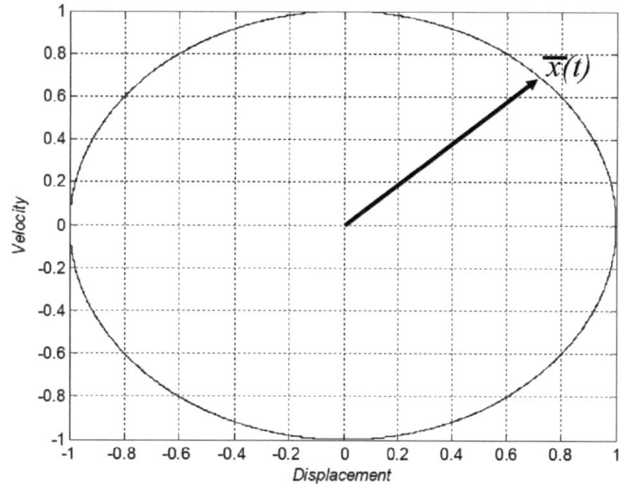

Figure 1.
Phase plane of a linear harmonic system.

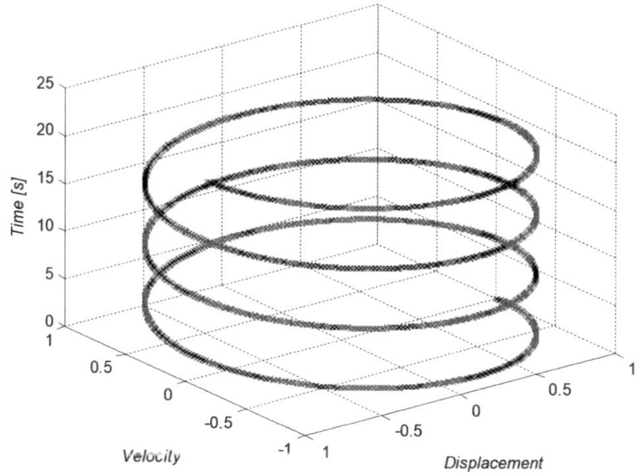

Figure 2.
Evolution of the state variables of a linear harmonic system.

$$R_{ij} = \begin{cases} 1 : \overline{x}_i = \overline{x}_j \\ 0 : \text{otherwise} \end{cases} \quad i,j = 1, \dots, N \tag{9}$$

The procedure applies to a continuous and discrete set of data. But, in general, data are discrete vectors; thus, N is the number of state vectors in the time array, divided by loops of period τ. Data contain noise and truncation errors; therefore, two vectors cannot have the same magnitude, and Eq. (9) is modified as:

$$R_{ij} = \begin{cases} 1 : \left|\overline{x}_i - \overline{x}_j\right| < \varepsilon, \\ 0 : \left|\overline{x}_i - \overline{x}_j\right| > \varepsilon, \end{cases} \quad i,j = 1, \dots, N \tag{10}$$

where ε is a tolerance value. The tolerance should be less than 10% of the mean diameter of the phase plane or five times larger than the standard deviation of the observational noise.

The orbits in a phase plane describe the system's dynamics. Orbits can see growth due to the entropy, and the number of orbits depends on the characteristic frequencies, damping, and the presence of nonlinear behavior. These characteristics modify the recurrence plots in a way that they can describe the system's behavior and predict future states. There are several characteristics of a recurrence plot that are classified as topology and texture. These characteristics are related to the system's dynamics.

Since the recurrence plots depend on the phase plane, it is important to construct a proper phase plane. The phase shift principle helps to build the phase plane of single-frequency signals; but for mechanical systems, this procedure introduces artificial noise that corrupts the phase plane [4]. The following sections describe the procedure for constructing the phase plane from acceleration measurements.

2.3 Topology and texture

In a Recurrence Plot, the main diagonal is always present, and parallel diagonals occur only in periodic or quasi-periodic systems. The distance between the diagonals is the fundamental period, and the diagonal lengths define if a system has a predictable behavior (steady condition). If the system presents recurrence but at different frequencies, then the diagonals will be shifted from the main diagonal.

Eckmann et al. [3] presented a summary of different topologies:

- A stationary system produces a homogeneous graph.
- If only some states change slowly or stop at a singularity, the graph has horizontal and vertical lines.
- Nonpersistent or fluctuating states produce isolated points. These points also mean that the states have continuous variations.
- Systems cause drift with slowly varying parameters; for example, nonstationary systems display a drift in the graphs.
- The main diagonal is interrupted when there is a sudden change, and the graph shows white areas.
- The graph shows dark and pale areas at the corners if the system has slow dynamic changes or slow varying parameters.

The topology depends on the tolerance, if ε is small, the Recurrence Plot erases most of the similar vectors, and it will be almost empty. Too large tolerance will produce many artificial figures that have no relation to the dynamic behavior.

2.4 Recurrence plot characteristics

Besides the topology and texture, a recurrence plot has several characteristics related to the system's dynamic response [39]. The following table (**Table 1**) summarizes the main features classified as recurrence quantification analysis.

The first step in constructing a recurrence plot is to build the phase plane. The problem with creating the phase plane using acceleration data is the integration of the

RQA	Equation	Description
Recurrence Rate	$RR = \frac{1}{N^2}\sum_{i,j}^{N} R_{ij}$	Measures the average number of recurrence points
Determinism	$DET = \frac{\sum_{l=l_{min}}^{N} lP(l)}{\sum_{l=1}^{N} lP(l)}$	Measures the percentage of recurrence points that lay on a diagonal. l is the index of the corresponding diagonal, $P(l)$ is the histogram of the diagonal. A chaotic system has short diagonals, and a periodic system forms regular diagonals, parallel to the main diagonal.
Longest Diagonal	$L_M = \max\{l_i\}$	Measures the length (in points) of the longest diagonal.
Longest Vertical Line	$V_M = \max\{v_i\}$	Measures the length of the longest vertical line.
Laminarity	$LAM = \frac{\sum_{v=v_{min}}^{N} vP(v)}{\sum_{l=1}^{N} vP(v)}$	Measures the percentage of recurrence points that create vertical lines. v is the index of the corresponding vertical line, $P(v)$ is the histogram of the vertical line
Shannon Entropy	$E = -\sum_{1}^{N} P_i Ln(P_i)$	Measures the disorder of the phase plane. For a linear and periodic system, the Shannon entropy is almost cero. For a non-deterministic and noisy system, the Shannon entropy will be almost infinite. $P_i = P(x_j = x_i)$ is the relative frequency, it is calculated as the ratio of diagonal with non-recurrence points divided by the number of recurrent points.

Table 1.
Recurrence quantification parameters.

state variables. The data are time series vectors $\ddot{x}$, and the phase plane is built with the state variables $\dot{x}$ and x. When the signal has only one frequency, the integration can be achieved by normalizing the original data and shifting them to one-quarter of the period. The following section describes the procedure for the integration of time series with any pattern.

3. Integration

The numerical integration has a fundamental problem because it integrates between limits, whereas for constructing the phase plane, the integration has to be indefinite. Torres and Jauregui demonstrated that numerical integrations introduce artificial errors in the phase plane [4]. They proposed a different integration procedure to avoid these errors that combine the one-quarter shifting method and the empirical mode decomposition (EMD) [5]. The EMD adapts the time domain to process nonstationary and nonlinear time series. The EMD method decomposes a signal into intrinsic mode functions by applying Hilbert's transform. The time series is approximated as [40].

$$a(t) = \sum_{1}^{n} IMF_i(t) + r(t) \tag{11}$$

The integration method begins by separating the acceleration signal into a set of intrinsic mode functions (IMF). The first iteration starts by identifying the points that determine the local maxima and minima. These sets of points define the lower and upper envelopes. Both groups of points are connected with a cubic spline. Then, the mean value function is the first IMF. The second step in the algorithm is to subtract the mean function from the original data. The following steps are finding the maxima and minima of the residual values, finding the mean function, and removing from the residual until the difference tends to minimum difference. The procedure is summarized in **Figure 3**.

Each intrinsic mode functions (IMF) is a smooth time series with variable amplitude and one or more frequencies, while the Fourier transform divides the acceleration signal into a set of harmonic functions with constant amplitude and frequency.

Once the acceleration data is divided into a set of IMF, the integration is based on the following principle.

Assume that the acceleration function is represented by:

$$a(t) = \mathrm{a}_1 \cos(\omega_1 t) + \mathrm{a}_2 \cos(\omega_2 t) + .. + \mathrm{a}_n \cos(\omega_n t) \tag{12}$$

The indefinite integral is:

$$\int a(t)dt = -\frac{\mathrm{a}_1}{\omega_1}\mathrm{sen}(\omega_1 t) - \frac{\mathrm{a}_2}{\omega_2}\mathrm{sen}(\omega_2 t) + .. - \frac{\mathrm{a}_n}{\omega_n}\mathrm{sen}(\omega_n t) \tag{13}$$

Or

$$\int a(t)dt = \frac{\mathrm{a}_1}{\omega_1^2}\cos(\omega_1 t + \tau_1) + \frac{\mathrm{a}_2}{\omega_2^2}\cos(\omega_2 t + \tau_2) + .. + \frac{\mathrm{a}_n}{\omega_n^2}\cos(\omega_n t + \tau_n) \tag{14}$$

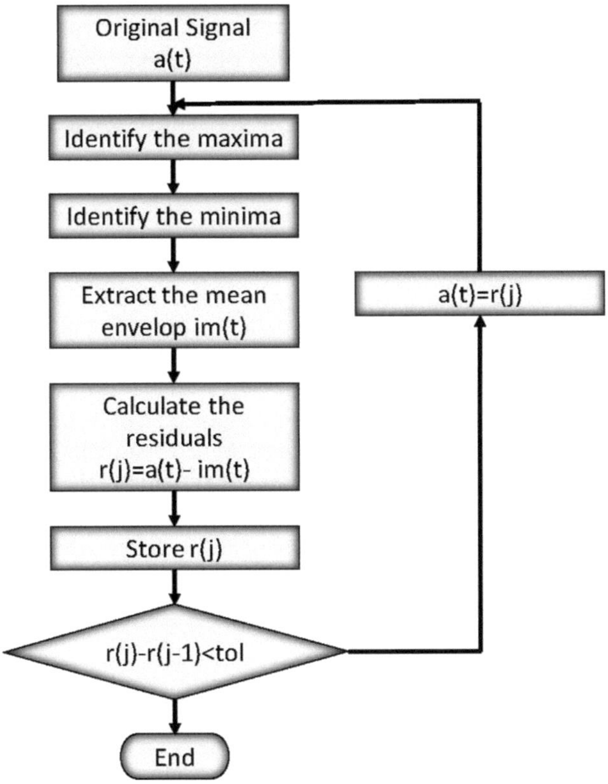

Figure 3.
EMD procedure.

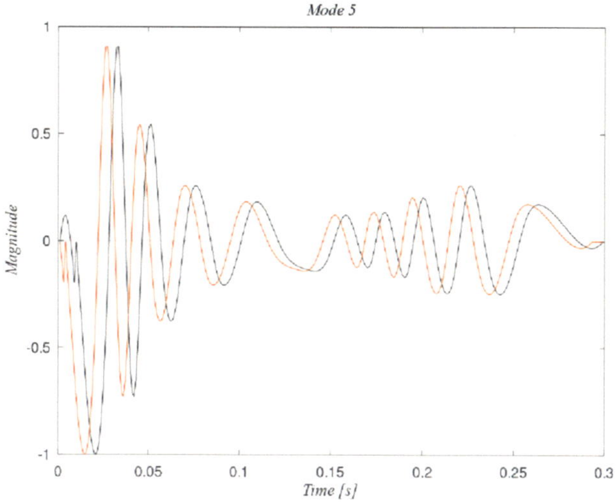

Figure 4.
Application of the shifting process to the intrinsic mode functions 5.

Combined with the EMD method, the shifting process (time delay) produces a better representation of the acceleration signal than the Fourier transform. **Figure 4** shows the application of the shifting procedure to one of the intrinsic mode functions (Mode 5). The integrated signal is obtained by adding all the shifted intrinsic mode functions.

Figure 5 shows the result of adding the shifted modes compared to the original signal.

The method based on the EMD and the shifting process was applied to measurements of a scaled-down experimental train.

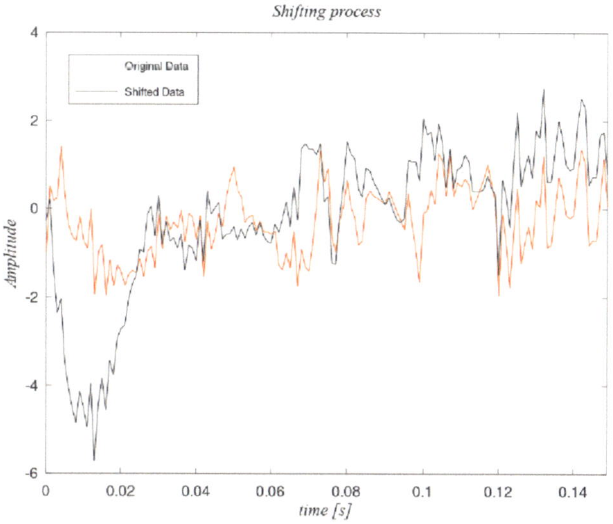

Figure 5.
Original and shifted signals produced by the EMD and shifting process.

4. Experiments

The scaled-down experimental railcar (**Figure 6**) was mounted on a closed-circuit track (**Figure 7**). The track has curves that make the railcar rotate clock and counter-clockwise. The railcar has two bogies, with a suspension and a platform that can be loaded with different weights and containers. One of the bogies is powered by a servomotor that can vary the railcar speed, and it is remotely controlled.

The railcar is instrumented with an encoder (for measuring the wheel's speed) with 500 PPR (pulses per revolution), three orthogonal accelerometers, and three gyroscopes (MEMs LSM6DS3) with a resolution of ±4 g and the gyroscopes ±143 DPS (degrees per second). As the railcar moved, a light sensor TCRT5000 counted the sleepers and determined the speed. The system had a data acquisition system with a 1000 Hz sampling rate.

The railcar ran freely around the track, but only the data from a segment was considered for the analysis. **Figure 8** shows the trajectory considered for the study.

The gyroscope data were used to locate the curvature changes within the data vectors. The Recurrence Plots can be built with any acceleration data; nevertheless, it was decided to use the vertical acceleration since it contains the higher dynamic responses and is more sensitive to tracking defects.

The following section describes the analysis of the measured data and the corresponding Recurrence Plots.

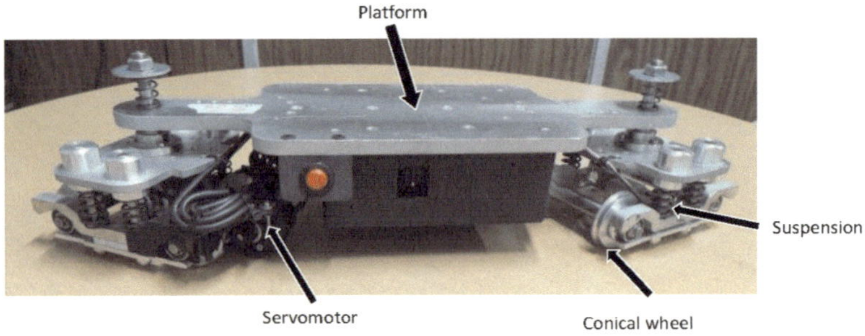

Figure 6.
Experimental railcar.

Figure 7.
Railcar mounted on the track.

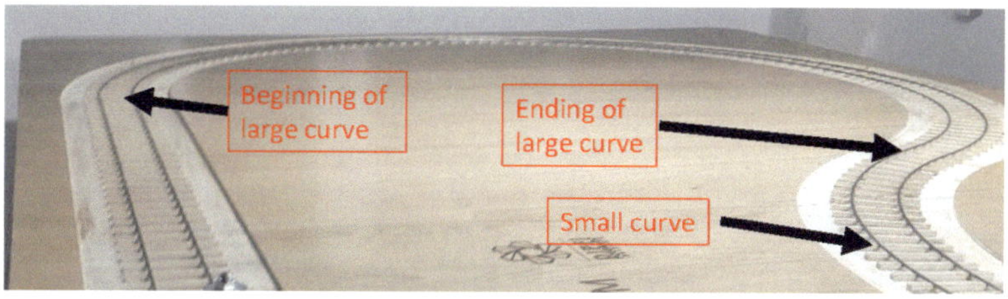

Figure 8.
Trajectory for data analysis.

5. Results

Figure 9 shows data obtained with the gyroscope (yaw rotation). This figure indicates the instants when the railcar entered the two types of curves. The data was segmented into short vectors whose length contained data generated when the railcar passed over an average of six sleepers. The data length is equivalent to the fundamental period of the vibration signals and gives a better representation of the dynamic behavior than more extensive data that can contain nosy values. The reason was to reduce the noise on the phase plane. The analysis consisted on:

1. Determining the IMF modes for the acceleration in the vertical direction (segmented vectors)

2. The IMFs were grouped into modes defining the rigid body and vibration motion modes. The rigid body motion modes were added to form a single signal named low frequency, and the vibration modes were added to create a signal called high frequency

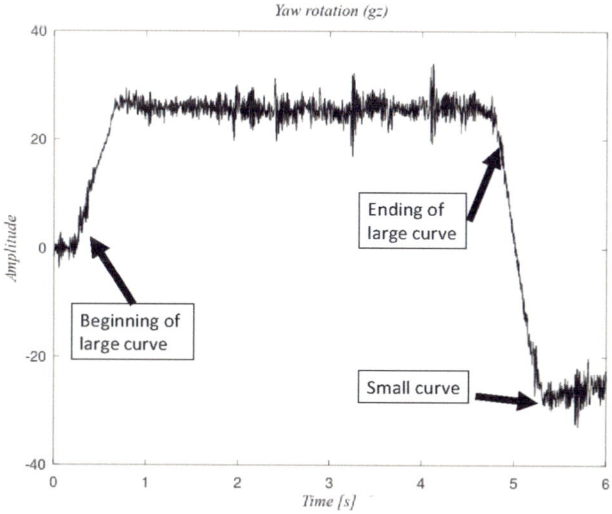

Figure 9.
Gyroscope measurements along the study segment.

3. The new signals were integrated following the shifting process (Eq. 14).

4. Constructing the Recurrence Plots for the low and high-frequency modes

The following paragraphs discuss the results based on the type of signal. The acceleration data were divided into the rigid motion modes, namely, low-frequency modes (intrinsic mode functions), and the vibration modes or high-frequency modes (intrinsic mode functions). The distinction came from the results obtained with the Empirical Mode Decomposition method.

Although the Recurrence Plots for produced for the entire trajectory, only the significant results are included in this chapter. **Table 2** provides the definition of the segments.

The first analysis corresponds to the curvature change from a straight segment into a large curve. **Figure 10** shows a picture of the track, identifying the beginning of the curvature change. The first sleeper in this segment has the tag "1." **Figure 15** describes the Recurrence Plots for the low- and high-frequency vectors. The low-frequency vector shows a high concentration of points along the diagonal and perpendicular diagonals, whereas the high-frequency vector presents diagonals and vertical and horizontal lines.

The following data correspond to a segment without defects (between sleepers 24 and 30). **Figure 11** shows a picture of this segment, and **Figure 16** presents the corresponding Recurrence Plots. In this case, both graphs show high concentrations along the main diagonal.

Segment	Description
1	Entrance to the first curve (**Figure 10**)
2	Segment without defects (**Figure 11**)
3	Joint defect (**Figure 12**)
4	Substructure defect (**Figure 13**)
5	Curvature change (**Figure 14**)

Table 2.
Description of the segments included in the results.

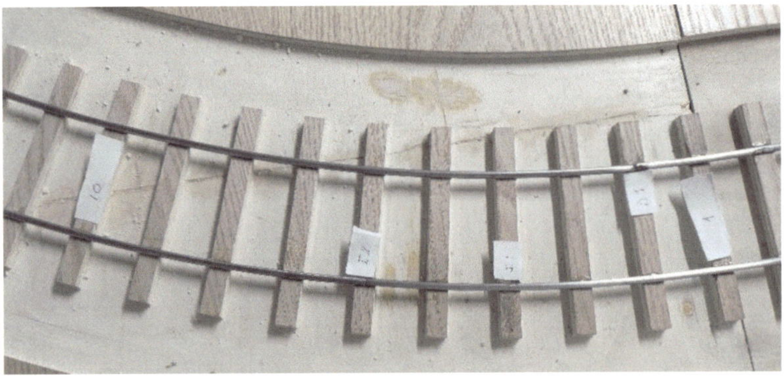

Figure 10.
Detail of the track at the beginning of the curve.

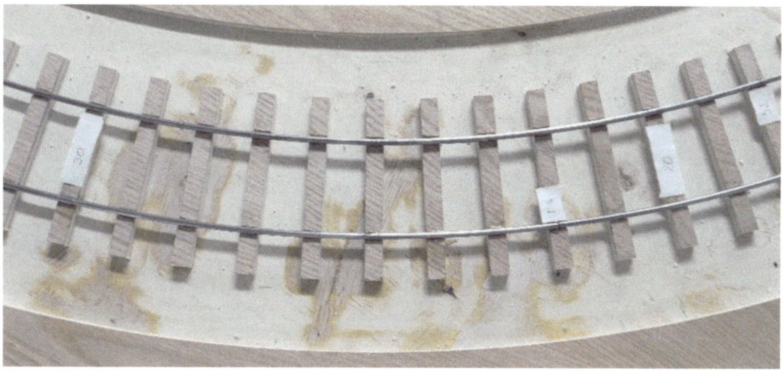

Figure 11.
Detail of the track in a segment without defects.

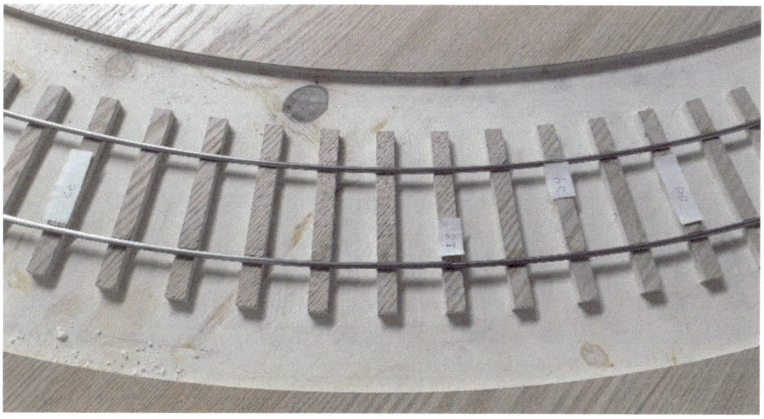

Figure 12.
Detail of the track with joint defect.

The defects on the track's joints produced impacts load on the railcar. These impacts occurred at several places along the track. An example of these defects is shown in **Figure 12**; the defects are located at numbers 52 and 54. The corresponding recurrence plots are presented in **Figure 17**. The low-frequency vector shows a concentration along the main diagonal; meanwhile, the high-frequency vector shows a horizontal and a vertical line that connects at a cluster of points in the main diagonal.

The discontinuity on the table that holds the track is assumed to be a substructure defect (**Figure 13** between sleeper 70 and 71). The dynamic response of the fault produced the recurrence plots shown in **Figure 18**. The low-frequency vector shows the main diagonal and a second perpendicular diagonal. The high-frequency vector shows perpendicular diagonals and a circular cluster of points that only appear in this plot.

Another dynamic condition related to the track design is the changes in curvature. Although the changes in curvature cannot be considered faults, they impose a different dynamic condition compared to straight or curved segments. **Figure 14** shows a picture of the curvature change; in this part, there are no joints or discontinuities in the track. **Figure 19** presents the recurrence plots when the railcar passed over this segment. The low-frequency vector has the main diagonal and a second perpendicular

Figure 13.
Detail of the track with joint and substructure defect.

Figure 14.
Detail of the track with a curvature change.

one with a large area in the intersegment. The high-frequency vector shows the main diagonal with a cluster of points at one extreme and two lines at the lower and upper corners. This pattern is different from other patterns along with the entire trajectory.

Other analyses can also predict or detect defects on the track. The Lyapunov exponent could be used to identify chaotic responses. The Fourier transform identifies the dominant frequencies; these analyses were considered for future work where the

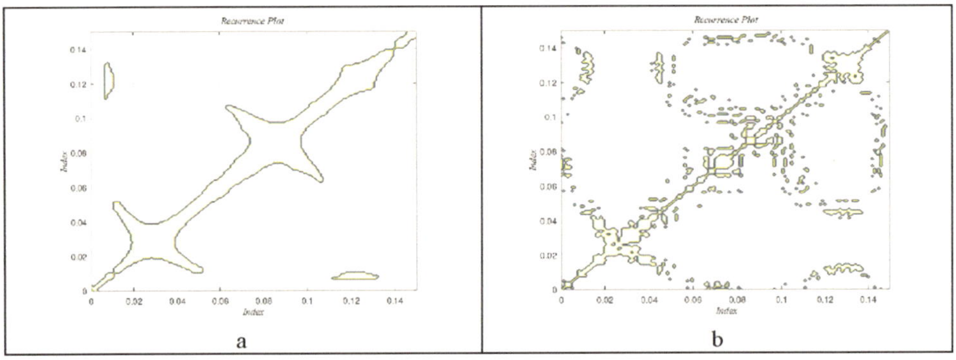

Figure 15.
Change of curvature, a) low-frequency modes (rigid motion), b) high-frequency modes (vibration modes).

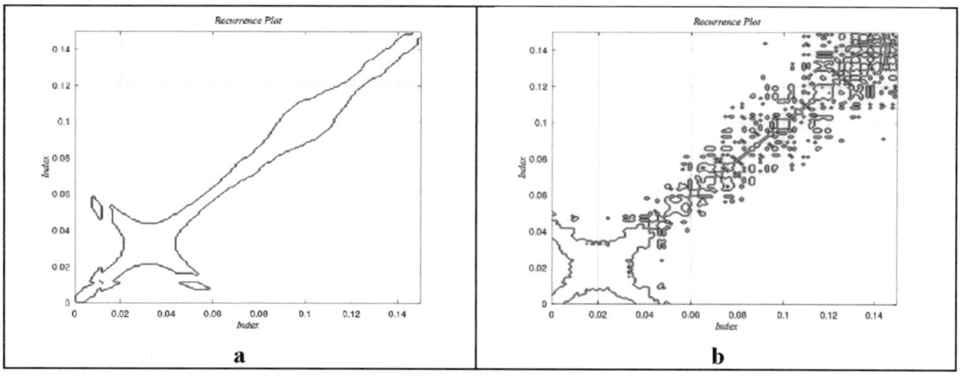

Figure 16.
Segment without defects, a) low-frequency modes (rigid motion), b) high-frequency modes (vibration modes).

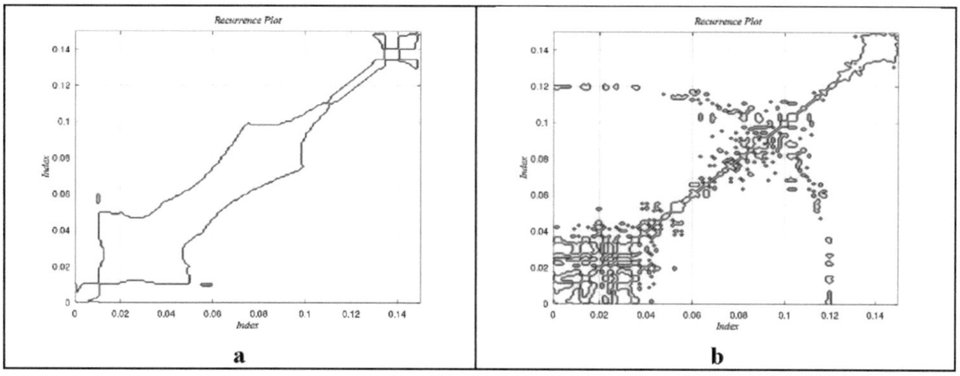

Figure 17.
Segment with a track discontinuity, a) low-frequency modes (rigid motion), b) high-frequency modes (vibration modes).

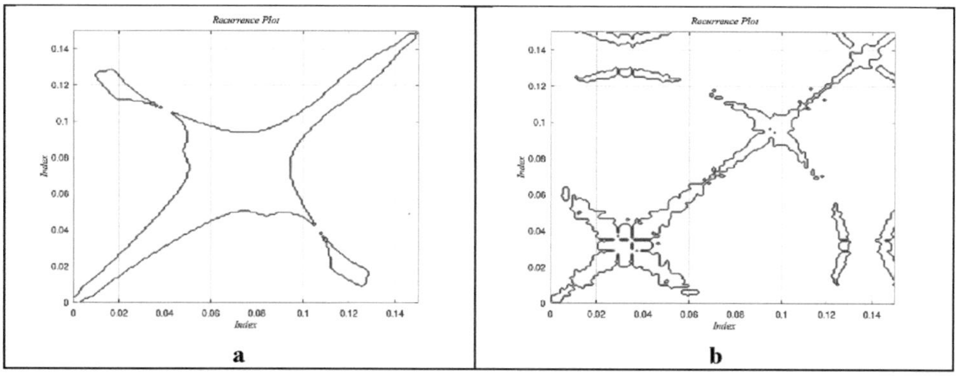

Figure 18.
Segment with a track and substructure defects, a) low-frequency modes (rigid motion), b) high-frequency modes (vibration modes).

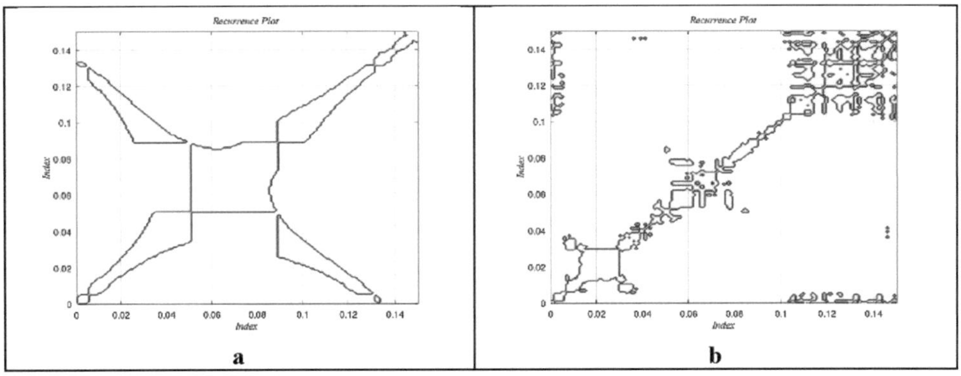

Figure 19.
Segment with a change in curvature, a) low-frequency modes (rigid motion), b) high-frequency modes (vibration modes).

Segment	Recurrence rate	Determinism (%)	Longest vertical	LAM (%)	Shannon entropy
1	1668	8.1342	24	16	0.011682
2	2262	10.792	39	26	−0.02885
3	1786	8.6622	25	16.667	−0.006406
4	1792	8.6890	27	18	−0.005191
5	1730	8.4116	26	17.333	0.015306

Table 3.
Recurrence quantification analysis (RQA) (high-frequency vector).

Segment	Recurrence rate	Determinism (%)	Longest vertical	LAM (%)	Shannon entropy
1	2602	12.313	35	23.333	0.0022262
2	2690	12.707	31	20.667	0.0036516
3	2128	10.192	28	18.667	−0.0004128
4	3154	14.783	40	26.667	0.0041652
5	3522	16.430	38	25.333	0.0059131

Table 4.
Recurrence quantification analysis (RQA) (low-frequency vector).

sensitivity of different analysis techniques could be compared. In this work, only the recurrence quantification analysis was used for comparison.

The recurrence quantification analysis was applied to the five segments. The analysis was applied to the high-frequency vectors since they showed the Recurrence Plots with higher variations. **Table 3** shows the results.

The same analysis for the low-frequency vectors is described in **Table 4**.

The RQA numerical values describe the dynamic system behavior. The Recurrence Plots obtained from the high-frequency showed more significant variations than the low-frequency data. Segment 2 data is a condition without multiple excitations; thus, it can be considered a reference. It had the lowest Shannon Entropy coefficient and the most considerable recurrence rate, determinism, longest vertical, and LAM.

The results presented in **Table 4** are more homogenous, except for Segment 3, which has lower values than the other segments.

The topology cannot be qualitatively analyzed. Thus, the recurrence plots need to be compared to each other. Further work will complement these results.

6. Conclusions

The application of recurrence plots to the dynamic analysis of vehicles, in this case, a scaled-down train, showed that variations on the track condition modified the graphical representation, and the recurrence quantification analysis demonstrated that these variations produced significant changes in the quantification parameters.

For producing the recurrence plots, acceleration measurements were recorded on the train's platform while traveling along a closed circuit. The vertical acceleration data were divided into the rigid motion modes (identified as low-frequency vectors) and the vibration modes (identified as high-frequency vectors).

The train's trajectory was divided into short segments to have cleaner graphs, approximately every six sleepers. This segmentation allowed the location of the defects within the time series (acceleration data). It was easier to locate the changes in curvature using the gyroscope data since it precisely showed the instants when the railcar changed direction and entered a curved section.

The recurrence plots were produced from the phase plane of each segment. The phase planes were calculated by separating the acceleration data into intrinsic mode functions using the empirical mode decomposition, applying the delay principle (wave shifting) to each intrinsic mode functions, and reconstructing the signal by adding them again. Before the addition, the intrinsic mode functions were grouped into two, one for the intrinsic mode functions associated with the rigid motion accelerations and the second for high-frequency vibration modes.

The high-frequency vector showed higher sensitivity to excitations. The track had several track joints, changes of curvature, and defects on the base plate. The recurrence plots showed different topologies that were considered "signatures" in all the cases.

The parameter that has a higher sensitivity for identifying defects on the track is the Shannon entropy. It measured the disorder on the recurrence matrices and provided a measurement tool for monitoring variation on the track or the railcar.

Further work should focus on automating the topology analysis and the definition of methodologies for predicting faults and defects using intelligent sensors on the railcar.

Nomenclature

$\boldsymbol{B}$	Matrix containing the system parameters
$\boldsymbol{x}$	Vector containing the state variables
T	Kinetic energy
p	Linear momentum$p = m\dot{q}$
V	Potential energy
q	Displacement
m	Mass

$\overline{x}(t)$	Vector of any point in the phase plane
R_{ij}	Element of the Recurrence Plot matrix
ε	Tolerance value
N	Number of points $N = \frac{n(n-1)}{2}$
n	Number of rows (or columns) of the Recurrence Plot matrix
a	Acceleration signal
t	Time intervals
r	Residuals
IMF_i	Intrinsic Mode Functions (IMF)

Author details

Juan Carlos Jauregui-Correa
Autonomous University of Queretaro, Queretaro, Mexico

*Address all correspondence to: jc.jauregui@uaq.mx

References

[1] Eckmann J, Oliffson S, Ruelle D. Recurrence plots of dynamical systems Europhysics Letters. 1987;**4**(9): 973-977

[2] Marwan N, Webber CL. Mathematical and computational foundations of recurrence quantifications. In: Marwan N, Webber CL, editors. Recurrence Quantification Analysis. Switzerland: Springer International Publishing; 2015. pp. 3-43

[3] Webber CL, Ioana C, Marwan N, Symposium I, Plots R. Recurrence plots and their quantifications: Expanding horizons. In: Proceedings of the 6th International Symposium on Recurrence Plots, Grenoble, France, 17-19 June 2015. Switzerland: Springer International Publishing; 2016

[4] Torres-Contreras I, Jáuregui-Correa JC, López-Cajún CS, Echeverría-Villagómez S. Effects of phase shift errors in recurrence plot for rotating machinery fault diagnosis. Applied Sciences. 2021;**11**. DOI: 10.3390/app11020873

[5] Torres-Contreras I, Jáuregui-Correa JC, Echeverría-Villagómez S, Benítez-Rangel JP, Camacho-Gutierrez S. Diagnosis of friction on an unbalanced rotor by phase-shift empirical mode decomposition integration and recurrence plot. Applied Sciences. 2021;**11**

[6] Joelle E, Bialonski S, Marwan N, Kurths J, Geier C, Lehnertz K. Evaluation of selected recurrence measures in discriminating pre-ictal and inter-ictal periods from epileptic EEG data. Physics Letters A. 2016;**380**(16): 1419-1425. DOI: 10.1016/j.physleta.2016.02.024

[7] Puthanmadam N, Hyttinen J. Characterization of dynamical systems under noise using recurrence networks : Application to simulated and EEG data. Physics Letters A. 2014;**378**(46): 3464-3474. DOI: 10.1016/j.physleta.2014.10.005

[8] Yan J, Wang Y, Ouyang G, Yu T, Li X. Using max entropy ratio of recurrence plot to measure electrocorticogram changes in epilepsy patients. Physica A. 2016;**443**:109-116. DOI: 10.1016/j.physa.2015.09.069

[9] Craciunescu T, Murari A. Geodesic distance on Gaussian manifolds for the robust identification of chaotic systems. Nonlinear Dynamics. 2016;**86**(1): 677-693. DOI: 10.1007/s11071-016-2915-x

[10] Saha D, Shaw PK, Ghosh S, Janaki MS, Iyengar ANS. Quantification of scaling exponent with Crossover type phenomena for different types of forcing in DC glow discharge plasma. Physica A. 2018;**490**:300-310. DOI: 10.1016/j.physa.2017.08.020

[11] Kwuimy CAK, Kadji HGE. Recurrence analysis and synchronization of oscillators with coexisting attractors. Physics Letters A. 2014;**378**(30–31): 2142-2150. DOI: 10.1016/j.physleta.2014.05.055

[12] Kwuimy CAK, Nataraj C. Recurrence and Joint Recurrence Analysis of Multiple Attractors Energy Harvesting System. Switzerland: Springer International Publishing; 2015. DOI: 10.1007/978-3-319-19851-4

[13] Kwuimy CAK, Nataraj SAAC. Characterization of the vibration transmitted in the human arm using

selected recurrence quantification parameters. Nonlinear Dynamics. 2017; **88**(4):2873-2887. DOI: 10.1007/s11071-017-3417-1

[14] Kwuimy C, Samadani M, Nataraj C. Bifurcation analysis of a nonlinear pendulum using recurrence and statistical methods: Applications to fault diagnostics. Nonlinear Dynamics. 2014; **76**:1963-1975. DOI: 10.1007/s11071-014-1261-0

[15] Jana D, Agrawal R, Kumar R. Top-predator interference and gestation delay as determinants of the dynamics of a realistic model food chain. Chaos, Solitons and Fractals. 2014;**69**:50-63. DOI: 10.1016/j.chaos.2014.09.001

[16] Thiel M, Romano M, Kurths J. Spurious structures in recurrence plots induced by embedding. Nonlinear Dynamics. 2006;**44**:299-305. DOI: 10.1007/s11071-006-2010-9

[17] Kurths J, Romano MC, Thiel M, Osipov GV, Ivanchenko MV, Kiss IZ, et al. Synchronization analysis of coupled noncoherent oscillators. Nonlinear Dynamics. 2006;**44**:135-149. DOI: 10.1007/s11071-006-1957-x

[18] Prakash J, Roy SBK. The simplest 4-D chaotic system with line of equilibria, chaotic 2-torus and 3-torus behaviour. Nonlinear Dynamics. 2017;**89**(3): 1845-1862. DOI: 10.1007/s11071-017-3556-4

[19] Litak G, Górski G, Mosdorf R, Rysak A. Study of dynamics of two-phase flow through a minichannel by means of recurrences. Mechanical Systems and Signal Processing. 2017;**89**: 48-57. DOI: 10.1016/j.ymssp.2016.08.037

[20] Zhai L, Zong Y, Wang H, Yan C, Gao Z. Characterization of flow pattern transitions for horizontal liquid – liquid pipe flows by using multi-scale distribution entropy in coupled 3D phase space. Physica A. 2017;**469**:136-147. DOI: 10.1016/j.physa.2016.11.020

[21] Xiong H, Shang P, Bian S. Detecting intrinsic dynamics of traffic flow with recurrence analysis and empirical mode decomposition. Physica A. 2017;**474**: 70-84. DOI: 10.1016/j.physa.2017.01.060

[22] Tang J, Wang Y, Wang H, Zhang S, Liu F. Dynamic analysis of traffic time series at different temporal scales: A complex networks approach. Physica A. 2014;**405**:303-315. DOI: 10.1016/j.physa.2014.03.038

[23] Vlahogianni EI, Karlaftis MG. Comparing traffic flow time-series under fine and adverse weather conditions using recurrence-based complexity measures. Nonlinear Dynamics. 2012;**69**:1949-1963. DOI: 10.1007/s11071-012-0399-x

[24] Ukherjee S, Ray R, Samanta R, Khondekar M. Nonlinearity and chaos in wireless network traffic. Chaos, Solitons and Fractals: The interd. 2017;**96**:23-29. DOI: 10.1016/j.chaos.2017.01.005

[25] Syta A, Litak G. Vibration analysis in cutting materials. In: Webber CL Jr, Marwan N, editors. Recurrence Quantification Analysis, Understanding Complex Systems. Switzerland: Springer International Publishing; 2015. pp. 279-290. DOI: 10.1007/978-3-319-07155-8_9

[26] Elias J, Namboothiri VNN. Cross-recurrence plot quantification analysis of input and output signals for the detection of chatter in turning. Nonlinear Dynamics. 2014;**76**:255-261. DOI: 10.1007/s11071-013-1124-0

[27] Leonardi G. A method for the computation of entropy in the recurrence quantification analysis of

categorical time series. Physica A. 2018; **512**:824-836. DOI: 10.1016/j.physa.2018.08.058

[28] Spiegel S, Schultz D, Marwan N. Approximate recurrence quantification analysis (aRQA) in code of best practice. In: Schultz D, Marwan N, editors. Recurrence Plots and Their Quantification: Expanding Horizons. Switzerland: Springer International Publishing; 2016. pp. 113-135

[29] Belaire-Franch J, Contreras D. Recurrence plots in nonlinear time series analysis: Free software. Journal of Statistical Software. 2002;7(9):1-18. DOI: 10.18637/jss.v007.i09

[30] Pham TD, Yan H. Spatial-dependence recurrence sample entropy. Physica A. 2018;**494**:581-590. DOI: 10.1016/j.physa.2017.12.015

[31] Girault J-M. Recurrence and symmetry of time series: Application to transition detection. Chaos, Solitons and Fractals. 2015;**77**:11-28. DOI: 10.1016/j.chaos.2015.04.010

[32] Sipers A, Borm P, Peeters R. Robust reconstruction of a signal from its unthresholded recurrence plot subject to disturbances. Physics Letters A. 2017; **381**(6):604-615. DOI: 10.1016/j.physleta.2016.12.028

[33] Le Bot O, Mars JI, Gervaise C. Similarity matrix analysis and divergence measures for statistical detection of unknown deterministic signals hidden in additive noise. Physics Letters A. 2015;**379**(40–41):2597-2609. DOI: 10.1016/j.physleta.2015.06.004

[34] Ngamkhanong C, Kaewunruen S, Alfonso-Costa B. State-of-the-art review of railway track. Infraestructures. 2018; **3**(1):1-18. DOI: 10.3390/infrastructures3010003

[35] Fermér M, Nielsen JCO. Vertical interaction between train and track with soft and stiff railpads—full-scale experiments and theory. Proceedings of the Institute of Mechanical Engineers. 1995;**209**(1):39-47. DOI: 10.1243/PIME_PROC_1995_209_253_02

[36] Ciotlaus M, Kollo G, Marusceac V, Orban Z. Rail-wheel interaction and its influence on rail and wheels wear. Procedia Manuf. 2019;**32**:895-900. DOI: 10.1016/j.promfg.2019.02.300

[37] Smith RA. Railway fatigue failures : An overview of a long standing problem. Material Science and Engineering Technology. 2005;**36**(11):697-705. DOI: 10.1002/mawe.200500939

[38] Jáuregui JC. Phase diagram analysis for predicting nonlinearities and transient responses. In: Baddourr N, editors. Recent Advances in Vibrations Analysis. London, UK: Intech; 2011. pp. 27-46

[39] Letellier C. Estimating the Shannon entropy: Recurrence plots versus symbolic dynamics. Physical Review Letters. 2006;**96**:55-60

[40] Cheng J, Yu D, Yang Y. A fault diagnosis approach for gears based on IMF AR model and SVM. EURASIP Journal of Advanced Signal Processing. 2008;**2008**. DOI: 10.1155/2008/647135

Chapter 4

Perspective Chapter: On Rolling Bearing Fault Feature Extraction Based on Entropy Feature

Yongjian Sun and Zihan Wang

Abstract

In large machinery, the most common element we can use is rolling bearing. When the rolling bearing fails, it is very likely to affect the normal operation of the equipment, or even cause danger. Therefore, it is necessary to monitor and diagnose the bearing fault in advance. The most important step in fault diagnosis is feature extraction. In this paper, the approximate entropy, the sample entropy, and the information entropy are analyzed, and the feature is extracted to diagnose the rolling bearing fault. Firstly, the concepts of approximate entropy, sample entropy, and information entropy are introduced briefly, and the approximate entropy, sample entropy and information entropy of rolling bearing vibration signals under different fault modes are calculated. The feasibility and shortcomings of the features extracted from these three entropy in the fault characteristics of rolling bearing are analyzed. In order to make up for their defects, a method of fault feature extraction based on approximate entropy, sample entropy, and information entropy is proposed, and its feasibility is verified. Simulation experiments are carried out to calculate the accuracy of fault feature extraction based on the joint analysis of approximate entropy, sample entropy, and information entropy.

Keywords: rolling bearing, fault diagnosis, approximate entropy, sample entropy, information entropy

1. Introduction

1.1 Research background and significance

In economic construction, as the main production equipment, it is inevitable that large-scale machinery will fail because of factors such as work difficulty or service time. However, the economic losses and casualties caused by the failure will make people lament [1]. In the 1990s, a power plant turbine unit caused great damage to the Korean high-speed rail derailment, causing many injuries and even deaths. Rolling bearing, as an important rotating part, has a long-term high-speed rotation, coupled with the precise and complex structure, and the failure may be quite high. Reliable data show that the motor faults caused by bearing faults are very frequent [2].

In the large or small machinery in various fields, the position of rolling bearings in it is indispensable and can be said to play an important role in production and life. Rolling bearing fault is the most common fault of rotating machinery, and once it happens, it will seriously affect the normal work of the whole machinery, so it is very important to study the fault diagnosis technology of rolling bearing [3–6]. For such an important rolling bearing, its fault diagnosis must be valued. Specific to the rolling body, inner circle, or external circle failures, we need to make subsequent improvements. If only one standard is conducted to detect and repair, not only the accuracy is low but also the cost of manpower and material resources [7]. If the fault diagnosis can be accurately conducted on prevention and moderate maintenance to avoid adverse effects. It will inevitably play a very important role in promoting the economic and social aspects [8].

1.2 Current status of fault diagnosis technology

Research on fault diagnosis of rolling bearings started around 1960. Overall, there can be divided into five stages [9].

The first stage is spectral analysis in the 1950s. Spectrum analysis methods have attracted attention. However, due to the immature technology at that time, the spectrum analysis was not widely used in the field of bearing fault diagnosis technology because the results were widely affected by the interference noise, expensive price, and complex operation.

In the second stage of the 1960s, the impact impulse meter detection method appeared, and the effect was obviously better than the spectral analysis, which could directly save the complicated steps, and was still widely used in the fault diagnosis of roller capital bearings.

The third stage in the 1960s to 1980s, computers and signals in the trend of The Times, the more prominent is resonance demodulation technology, because of the advent of this technology, makes the rolling bearing fault diagnosis technology to a higher level, gradually from the beginning to maturity.

After the 1980s, the emergence of artificial intelligence provided a new soil for the rolling bearing fault diagnosis and the emergence of an intelligent diagnosis system greatly improved the accuracy of the fault diagnosis. Due to intelligence, the influence of human factors is greatly reduced, which has been applied in engineering practice.

The fifth stage is after the 21st century, that is, we are now, the rolling bearing fault diagnosis technology has taken an epoch-making step, more and more high-tech development, through the fault diagnosis of virtual instruments, has become a new pointing mark, has an important practical value.

At present, around the world, we are constantly studying the rolling bearing fault diagnosis, using a large number of different research fields. According to the most popular classification methods, it is divided into three categories, namely, model-based fault diagnosis technology, knowledge-based fault diagnosis technology, and data-based fault diagnosis technology.

Because of its national conditions, China began to study fault diagnosis much later than that in other countries. In the late 1970s, it was not first available in the early 1980s and formal research began. However, it is gratifying that under the hard work of Chinese researchers, in the 1990s, the field of fault research has been on the right track, and it has made great breakthroughs in both theory and practice and can be applied in production and life. But compared with other countries, China still has a long way to go.

1.3 Main work of the paper

In this paper, we study joint analysis of rolling bearings based on approximate entropy, sample, and information entropy.

It roughly describes the historical background and practical significance of fault diagnosis research in the 21st century today and briefly expounds on the current situation of the global research on fault diagnosis.

The universal structure of the rolling bearing is introduced and the relevant parameters are marked in the plane structure diagram. The most common form of rolling bearing failure is described, and the characteristics and hazards of the bearing are also mentioned.

The cause and mechanism of the bearing vibration are expounded in detail. A theoretical method for calculating the characteristic frequency of the rolling bearing is presented. Bearing fault diagnosis experimental equipment (Western Reserve University) is introduced, and its relevant basic parameters and bearing fault setting form are introduced.

Concept definitions and calculations for approximate entropy sample and information entropy are given. A single entropy feature of the rolling bearing vibration signal in different failure modes is extracted, combined with data to verify the feasibility of approximate entropy and sample entropy in terms of failure features. The three entropy features are jointly analyzed to distinguish the different fault modes of bearings.

In the end, the full research work is summarized, the deficiencies are proposed, and the future research is preliminary.

2. Basic nature of the rolling bearing

2.1 Structure and failure forms

2.1.1 Basic structure and parameters

The most basic structure of the rolling bearing is the outer ring in the outside, and the inner ring inside, and in the middle through the rolling body like the ball or the column, so as to transform the sliding friction into the rolling friction. In general, the outer ring is fixed, and the inner ring is connected with the matched axis, and the holder is also a common structure of the rolling bearing, which can avoid the friction seen by the rolling body.

The quality of the rolling body is quite critical, and a good rolling body can well reduce the friction coefficient, reduce the friction, loss, and the bearing rotation will become smooth and efficient. The smoother the inner circle is, the less the friction is in contact with the axis, and the smoother the rotation will be. The outer ring plays a supporting role, hardness must be high, otherwise easy to wear, cause bearing off. The role of the holder is to fix the rolling body; the size must fit with the rolling body, too large and too small will cause adverse effects.

As shown in **Figure 1**, we can clearly see the structure of the rolling bearing through the 3-dimensional diagram, and the main parameters are marked in the plane structure diagram, including: inner circle radius r_1 , outer circle radius r_2 , rolling body diameter d, bearing joint diameter D, rolling body number Z, bearing contact angle θ.

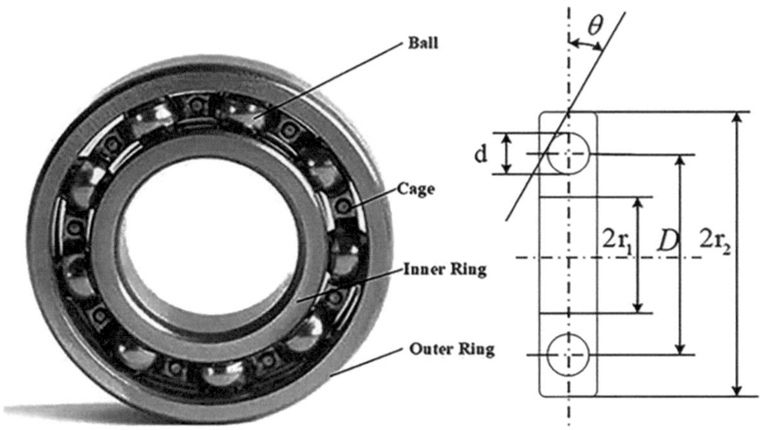

Figure 1.
Drawing and plane structure of the rolling bearing.

2.1.2 Common failure forms

Rolling bearing, as a component part of the mechanical equipment, will inevitably produce a loss. Even if their own quality, including material processing and other aspects are very excellent, and there is no working condition failure, after a period of use, there will still be fatigue and wear, and other conditions. Master the common failure form of bearings, and can repair and replace the bearing elements in advance to extend the service life [10].

1. Fatigue peeling

Under the repeated action of various forces, rolling bearing outer ring and inner ring, the maximum force part appear crack, surface metal may show point peeling, in serious cases even sheet peeling, this phenomenon, we call fatigue peeling. Fatigue peeling is the most common without special circumstances.

2. Surface plastic deformation

The surface of the rolling bearing will form mechanical damage because of the action of pressure, or hard objects involved in the rolling bearing, and the damage continues to expand, which will cause vibration, noise, etc. so that the damage speed of the bearing is greatly accelerated. This phenomenon is surface plastic deformation.

3. Corrosion

When the rolling bearing is working, the surface and internal metal and the substances in the environment, such as acid and alkali, or the consumption phenomenon caused by chemical reaction into water, that is corrosion.

4. Wear and tear

The relative movement of the two contact metal surfaces is inevitable. In relative motion, friction, metal consumption, deformation, change the size of the rolling bearing, and then cause a change in performance. This phenomenon is wearing.

5. Wriggle

Rolling bearing in the work, affected by the load, the inner ring and axis in rotation, in the circular direction, relative movement, on the metal surface of friction, wear and other abnormal damage, this phenomenon is peristalsis.

6. Scaling loss

In the process of use of the rolling bearing, due to the excessive bad lubrication load and other factors, itself is affected by high temperature, and not timely cooling will make the element surface burn, serious, the probability of the rolling bearing stuck, this phenomenon is burning loss.

2.1.3 Category of rolling bearings

The most used part in the machinery industry is bearings, and because bearings are often needed in all walks of life, the categories of bearings are also very diverse. The bearing can be divided into multiple categories according to the rolling body shape, column number, and outer diameter size of the bearing [11].

According to the shape of the rolling body, it can be divided into the ball bearing and the ball bearing, the ball bearing is well understood, that is, the rolling body is the ball bearing, and the rolling body in the roller bearing is generally other types of rolling body such as the cylindrical roller.

If the rolling bearings are classified according to the bearing load direction, they can be divided into thrust bearings, centripetal bearing, and centripetal thrust bearing. The load direction of the thrust bearing is from the axial direction, while the center bearing mainly bears the radial load. And the centripetal thrust bearing is more powerful, can bear the load formed by the axial and radial combination, but because of this reason, the life of the centripetal thrust bearings is often shorter than other bearings.

At the same time, it can also be classified according to the number of rolling body columns, there are single column bearings, double column bearings, and multiple column bearings, the literal meaning can be understood, and it will not repeat.

Finally, it can also be classified by the size of the bearing outer diameter, from micro bearings less than 26 mm in diameter to a major class bearing greater than 2000 mm in diameter, divided into bearings of various specifications.

2.2 Vibration mechanism

1. Vibration caused by the structural characteristics

Inherent vibration of the collar, vibration caused by the elastic characteristics of the bearing, vibration when the rolling body passes through the bearing area.

As a mechanical element, vibration is inevitable, mainly reflected in the outer circle, and is determined by its own structural characteristics. When the rolling bearing is affected by the external force, the external circle vibration inevitably occurs.

Rolling bearing in work, it is impossible to idling, must bear the load. Usually, the load is not small, which requires the rolling body to be very rigid. But large rigidity means that in special cases, the rolling experience produces a spring-like effect, producing vibration.

The different number of rolling bodies serving as supports when passing through the carrying area also causes the inner circle ring to vibrate back and forth in the front and rear directions.

2. Vibration caused by the rolling bearing processing process

In the processing process of the bearing production, due to the equipment accuracy and other problems, there will inevitably be an inner ring, and the outer ring will have a slight fluctuation. Only in the high-speed rotation, the effects of these fluctuations are also magnified, causing vibration.

On the other hand, the uneven size of the rolling body is also one of the causes of the vibration, which will greatly reduce the service life of the bearing.

3. Vibration caused by component failure

When the failure of the outer ring or the rolling body in the outer ring of the inner circle occurs, the vibration situation is also different. Different data can be obtained.

2.3 Feature frequency

To quantify the processing, the bearing fault characteristic frequency is given below. The rotation frequency of the ordering axis is f_r, the number of rolls is z, d is the diameter of ball, D is the diameter of rolling frame, and contact angle θ. When the outer ring of the rolling bearing is fixed, the theoretical calculation formula of the fault characteristic frequency of each element is as follows [12].

Inner ring fault frequency:

$$f_i = \frac{Zf_r}{2}\left(1 - \frac{d}{D}\cos\theta\right) \tag{1}$$

Outer ring fault frequency:

$$f_o = \frac{Zf_r}{2}\left(1 + \frac{d}{D}\cos\theta\right) \tag{2}$$

Ball failure frequency:

$$f_{BS} = \frac{Df_r}{2d}\left(1 - \left(\frac{d}{D}\right)^2 \cos^2\theta\right) \tag{3}$$

2.4 Diagnosis test of rolling bearing

2.4.1 Test device

Figure 2 shows the equipment used in the bearing data center of the rolling bearing data, with a 1kw motor on the left as the power provider, while in the center is the torque sensor, on the right is the power measuring motor, and the electronic control device.

Figure 2.
Experimental Equipment of Western Reserve University.

2.4.2 Basic parameters of rolling bearing fault

In this experiment, 6205-2RSJEMSKF bearing was used. The fault setting of the bearing was a single fault, and four damage degree faults were set on the inner ring, outer ring, and rolling body, respectively, namely 0.1778 mm, 0.3556 mm, 0.5334 mm, and 0.7112 mm. The rotational speed of the motor is 1797 r/min, 1772 r/min, 1750 r/min, and 1730 r/min, respectively.

This paper uses the SKF6205 bearing at a 12K sampling frequency, a fault diameter of 0.1778mm, a motor speed of 1750 r/min, and the vibration signal from the drive end bearing fault data of the acceleration sensor at the 6 o'clock position.

2.5 Influence and relationship between rolling bearing motion and chaos

2.5.1 Chaos phenomenon

With the advent of Lorentz nonlinear dynamical system, chaos has attracted more and more attention. As the representatives of nonlinear dynamical systems, Lorentz equation, and Lorentz-like equation have attracted the general attention of many scholars at home and abroad, and they have been deeply studied.

Lorentz system is a pioneer of chaos research. Chaos research based on Lorentz system can be divided into two independent methods, one is the study of the properties of equation solutions, numerical simulation by computer [13–15], and the other is chaotic water wheel physics experiment.

In Lorentz nonlinear dynamical system, the degree of randomness is generally determined by entropy. At present, in Lorentz nonlinear dynamical system, the determined entropy has no name, so we can call it Lorentz entropy for the time being.

The overall chaotic level of the system can be measured by the maximum Lyapunov exponent, which quantitatively describes the divergence rate of the phase volume exponent of the adjacent orbits of the system in the phase space. Although chaos is an irregular phenomenon, it comes from the deterministic system, so it is possible to predict it in a short term.

For Lorentz nonlinear dynamical systems, there is a relationship between Lorentz entropy and Lyapunov exponent. In chaotic one-dimensional mapping, a single Lyapunov exponent is consistent with Lorentz entropy.

An important quantitative method to judge whether the system is chaotic is whether there is a positive Lyapunov exponent. Lyapunov exponent is an important quantitative index reflecting the characteristics of dynamic system, which indicates the long-term average exponent of convergence or divergence between adjacent orbits of the system in phase space. For a time-delay dynamic system, its initial condition is a continuous function, so its Lyapunov exponent is related to the continuous function as the initial condition. The continuous function defined in the initial time period is uncountable, and so is the number of Lyapunov exponents of the system. Calculating Lyapunov exponents of time-delay dynamical systems is a complicated task. In the process of calculation, there may be some strange situations that make the results inaccurate. Therefore, judging whether the system is chaotic, as long as the maximum Lyapunov exponent is greater than zero, it can be used as a reliable basis for the existence of chaos.

The following diagram shows the transformation of the maximum Lyapunov exponent with parameters in the chaotic waterwheel experiment of Lorenz typical system. It can be seen that when the parameter reaches 15, the maximum Lyapunov exponent is positive, resulting in chaos.

2.5.2 Brief introduction of chaotic waterwheel device

As shown in **Figure 3**, the chaotic waterwheel device is similar to the ancient waterwheel, with a constant water flow at the top of the waterwheel injected into the water cup hanging on the edge of the wheel. There is a small hole at the bottom of each cup that can constantly discharge water. If the water flow speed on the top is very slow, the water in the top cup is small, so the friction force of the axle cannot be overcome, and the water wheel will not rotate; If the water flow speeds up, with the increase of water in the top cup, the water wheel will start to rotate at a constant speed; As the water flow continues to increase, the rotation will be chaotic, and the direction and speed of rotation will have complex motion characteristics due to the inherent nonlinearity of the system (**Figure 4**).

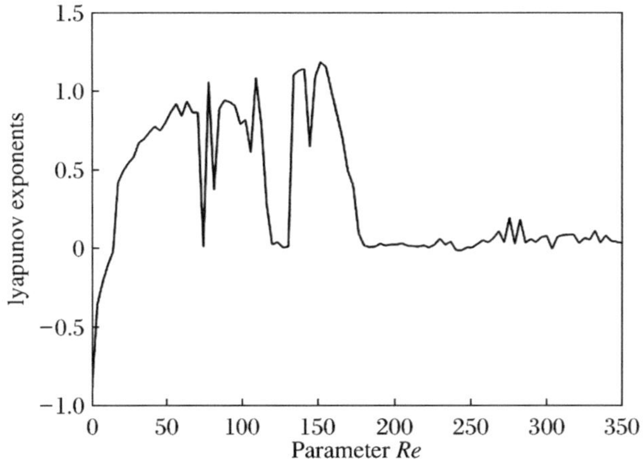

Figure 3.
Maximum Lyapnov exponential.

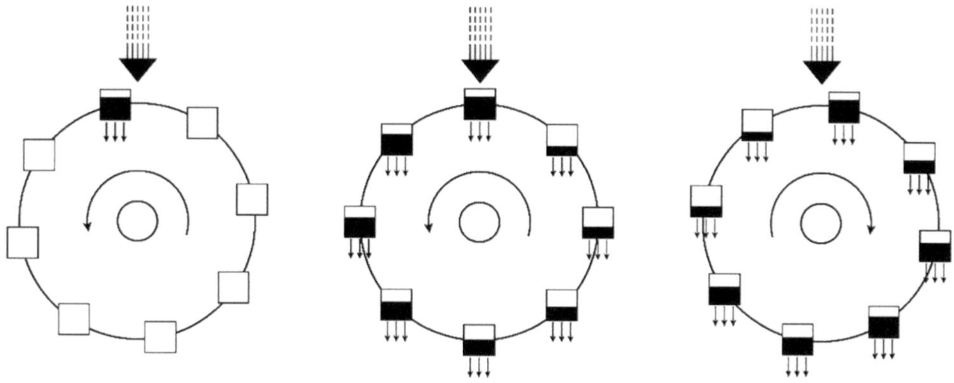

Figure 4.
Schematic diagram of chaotic water wheel experimental device.

2.5.3 Influence and relationship between rolling bearing motion and chaos

Rolling bearing is similar to chaotic water wheel device. The fault location will affect the rotation of rolling bearing, and the rotation will be chaotic. The direction and speed of rotation will have complex but regular motion characteristics due to the inherent nonlinearity of the system. Because of the different fault locations, the vibration signals presented by the rolling bearing rotation are different.

2.5.4 Entropy and chaos phenomenon

Entropy is a measure of disorder, in a chaotic system. The greater the entropy, the more chaotic the chaotic system is. The smaller the entropy, the more orderly the chaotic system is. Approximation entropy, sample entropy, and information entropy are all entropy, and as characteristics, they also affect the chaotic degree of chaotic phenomena.

2.6 Summary

The basic structure of the rolling bearing and the relevant parameters are introduced briefly. Six common rolling bearing failures are subsequently described. The cause of bearing vibration is clearly explained and the method of the characteristic frequency of inner ring, outer ring, and rolling body is given. We briefly introduce the western storage university bearing fault diagnosis equipment and state the data used in this paper.

3. Analysis of rolling bearing fault features based on approximate entropy, sample entropy, and information entropy

3.1 Nonlinear dynamic analysis of vibration signal

3.1.1 Time domain analysis

Figures 5–8 are the time domain signals of four working conditions of rolling bearing in turn.

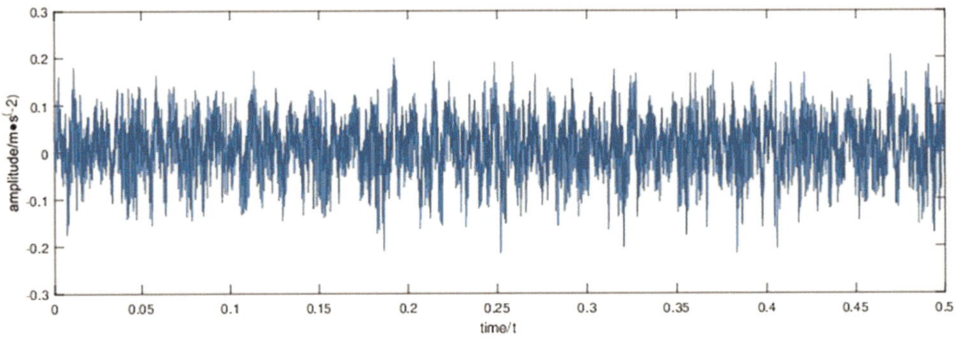

Figure 5.
The normal bearing signal.

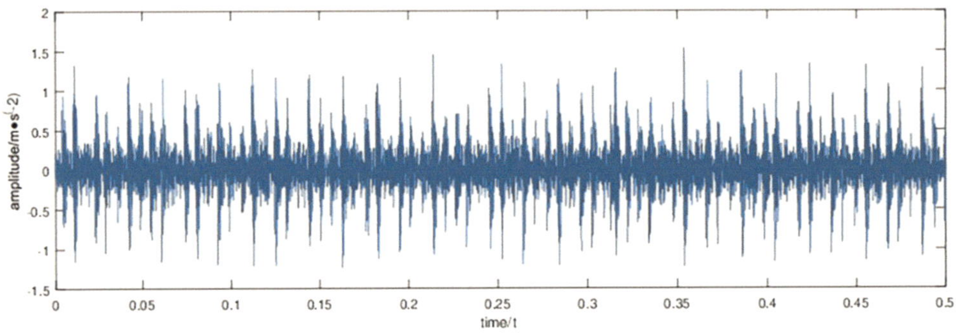

Figure 6.
The inner ring fault signal.

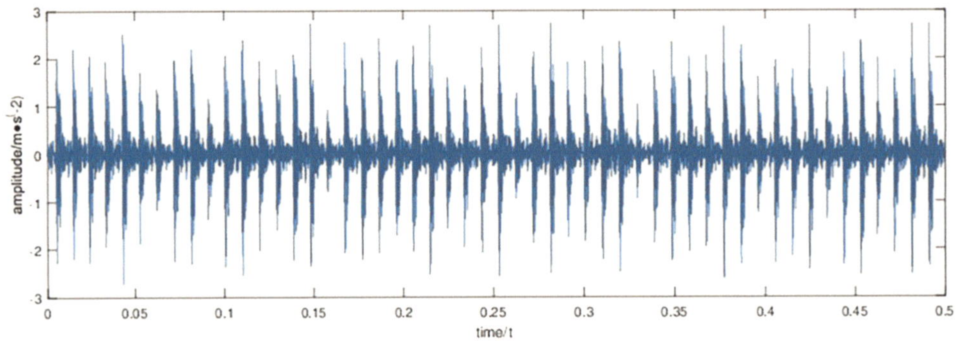

Figure 7.
The outer ring fault signal.

From the time domain image of the vibration signal of the rolling bearing, it can be seen that the vibration signals generated by the bearing under different working conditions are different, although there are differences, but they are not obvious. No matter whether the human eye or the computer can make an accurate judgment only by the vibration signal, the error is great. Therefore, it is necessary to dig deep into the vibration signal data and find out the characteristics of each working condition.

By calculating the entropy of the vibration signal of rolling bearing, its characteristics can be extracted well.

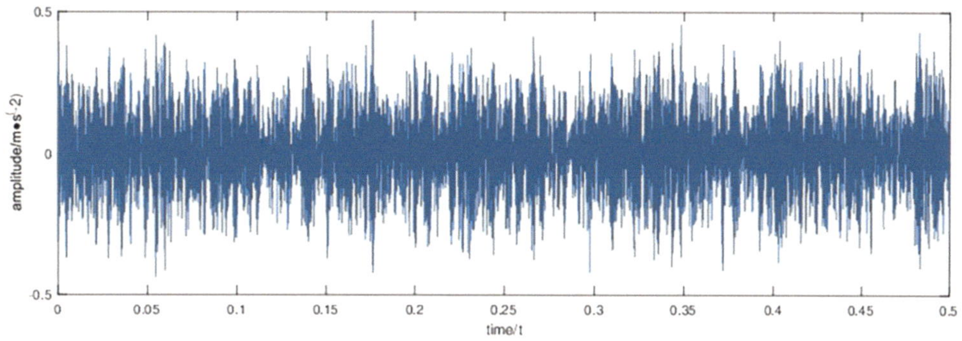

Figure 8.
The rolling body fault signal.

3.1.2 Fast Fourier Transform of power density spectrum

Fast Fourier transform (FFT), that is, the general name of the efficient and fast computing method of computing discrete Fourier transform (DFT) by computer, is abbreviated as FFT. Fast Fourier transform was proposed by J.W. Cooley and T.W. Tuki in 1965. By using this algorithm, the number of multiplications required by computer to calculate discrete Fourier transform can be greatly reduced. Especially, the more sampling points N are transformed, the more significant the saving of FFT algorithm is.

Fast Fourier transform (FFT) is a method to quickly calculate the discrete Fourier transform or its inverse transform of a sequence. Fourier analysis can convert the signal from the original domain (usually time or space) to the frequency domain for representation. For sequence $x(n) = \{x_0, x_1, \dots, x_{N-1}\}, 0 \leq n < N$, the discrete Fourier transform expression is:

$$\hat{x}(k) = \sum_{n=0}^{N-1} x(n) e^{-i\frac{2\pi}{N}nk} \tag{4}$$

Figures 9–12 are the Fast Fourier Transform of vibration signals of the rolling bearing in four working conditions.

It can be seen that the normal working conditions have obvious peaks at frequencies of 0,500 and 1000Hz, which are very regular. When the inner ring fails, the peak value is concentrated at 1000~2000Hz. When the rolling element fails, the peak value is concentrated at 1500~1800Hz. When the outer ring fails, the peak values are concentrated in 800~1500Hz and 1700~1800Hz. Therefore, FFT algorithm can not only diagnose signals without noise, but also apply to fault signals with noise.

Figures 13–16 are the calculation of the power spectrum function of vibration signals of the rolling bearing in four working conditions.

Through the power density spectrum, it can be seen that the power density distribution range of vibration signals of rolling bearings in different working conditions is different, which means that the chaotic degree of this nonlinear system is different, and the entropy is naturally different, so it can be extracted as a feature.

The collected vibration signal is greatly influenced by noise, so the time domain signal cannot be directly extracted. From the analysis of the frequency domain curve, it can be seen that there are peaks at the characteristic frequencies of each working

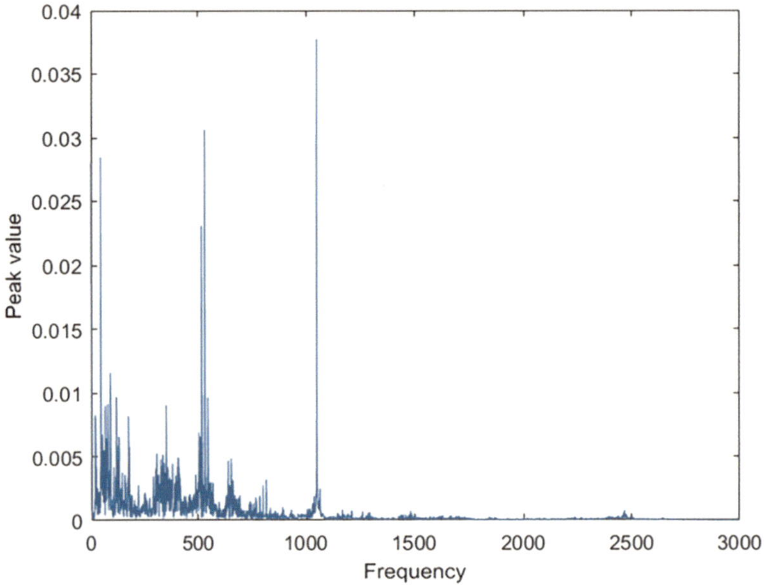

Figure 9.
Normal Working Condition Fast Fourier Transform.

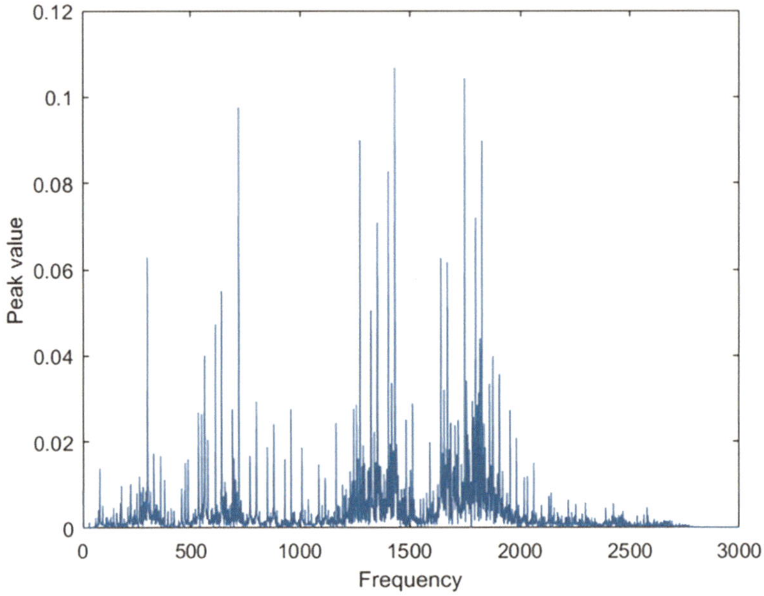

Figure 10.
Inner Ring Fault Fast Fourier Transform.

condition. Under normal working conditions, the peak amplitude obtained by fast Fourier transform is concentrated at a certain frequency. However, the vibration data of the inner ring fault, outer ring fault, and rolling element fault, and the amplitude peaks obtained by fast Fourier transform are scattered at various frequencies, showing different chaotic phenomena and different chaotic degrees.

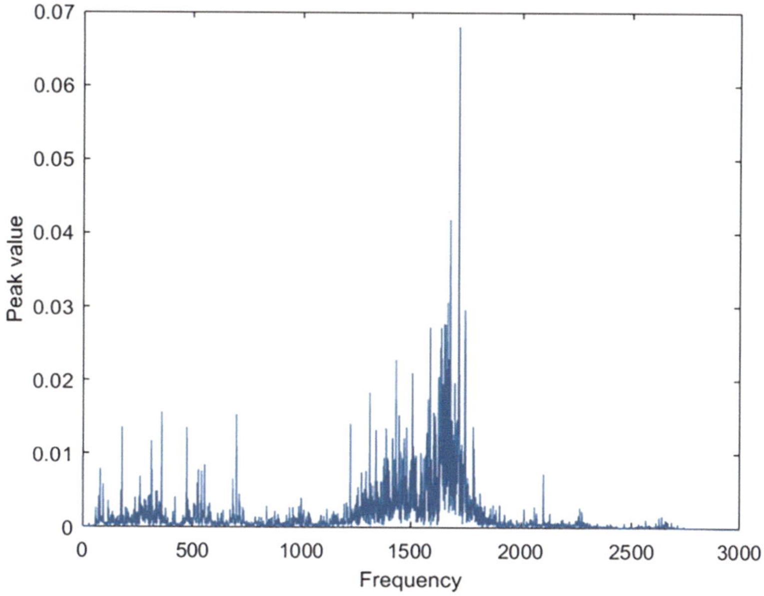

Figure 11.
Rolling Element Fault Fast Fourier Transform.

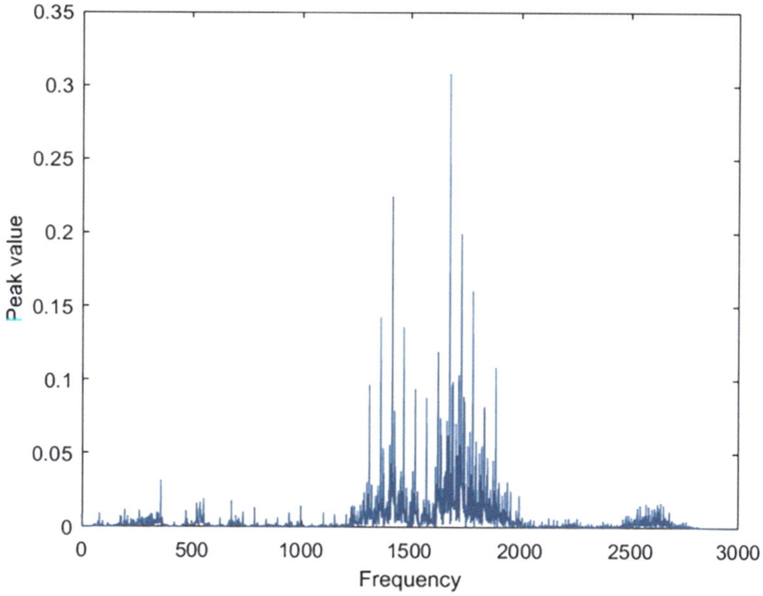

Figure 12.
Outer Ring Fault Fast Fourier Transform.

However, the frequency components such as frequency conversion and frequency doubling are not obvious in the frequency domain curve. If the fault degree is light or the fault mode is complex, the characteristic frequency peak of each working condition is likely to be submerged in the noise, and it cannot be identified by the frequency domain diagram alone. However, different chaotic degrees mean different entropy, so entropy can be used as the fault characteristics of rolling bearings.

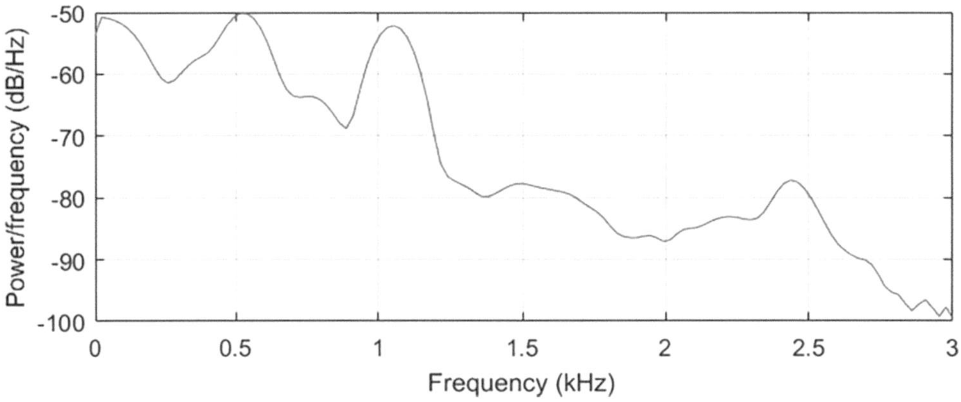

Figure 13.
Normal Working Condition Power Spectral Density Estimate.

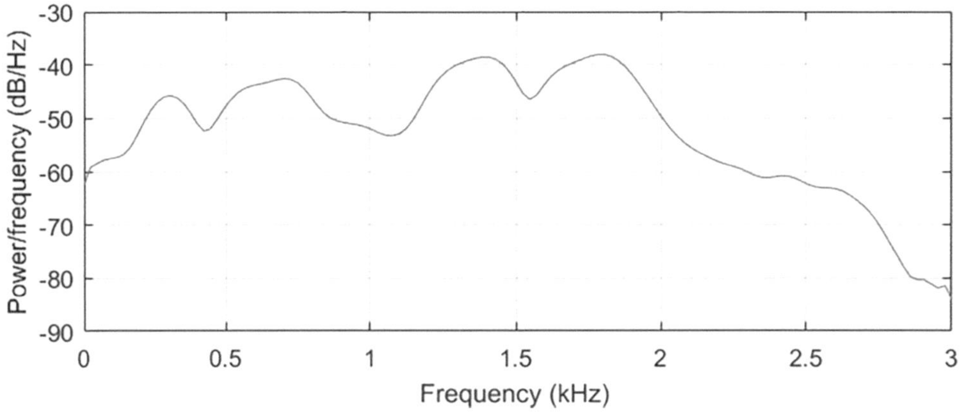

Figure 14.
Inner Ring Fault Power Spectral Density Estimate.

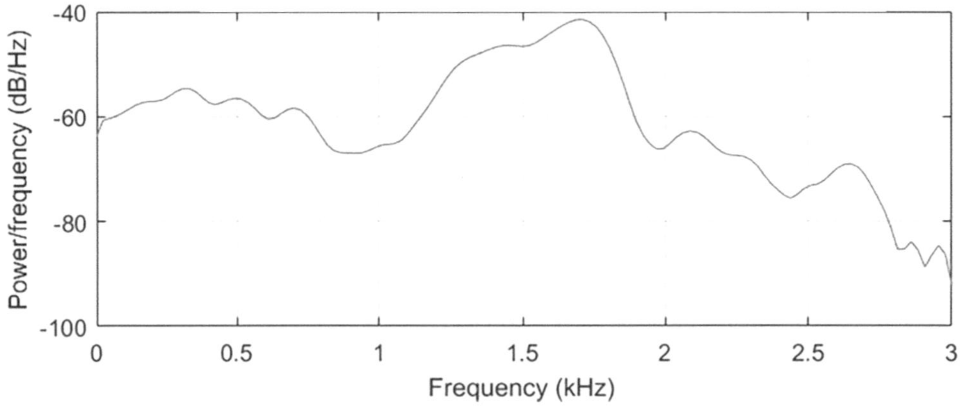

Figure 15.
Rolling Element Fault Power Spectral Density Estimate.

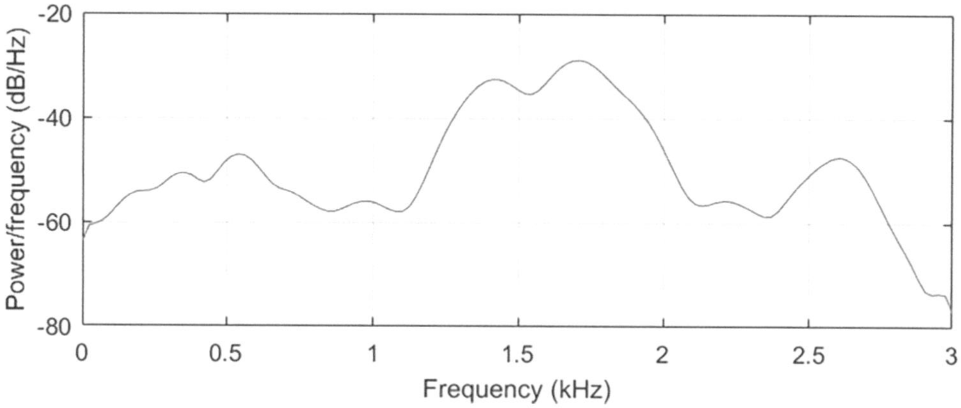

Figure 16.
Outer Ring Fault Power Spectral Density Estimate.

3.2 Approximate entropy

3.2.1 The concept of the approximate entropy

Approximate entropy ApEn is a non-linear kinetic parameter of sequence proposed by Pincus in 1991. ApEn reflects the degree of self-similarity of the sequence in the pattern [16].

The larger the ApEn value means that it is a complex sequence, and the less likely the system will be able to predict it. It gives cases where the incidence of new patterns increases with dimension, thus reflecting the structural complexity of the data.

From the above, we can know that the rolling bearing produces vibration, and the vibration signal are different in different failure modes. Depending on the physical meaning of ApEn, different signals imply different complexities that can be used as features for rolling bearing fault diagnosis.

3.2.2 Fast algorithm for the approximate entropy

When calculating approximate entropy, too much extra calculation is a waste of time. A fast algorithm to approximate the entropy is presented, exactly as follows:

Let the original sequence be {u (i), i = 0,1, ..., N}, r = 0.1~0.25SD (u) (SD indicates the standard deviation of the sequence {u (i)}), then the approximate entropy is more reasonable, select m = 2, N = 500–1000.

Calculated distance matrix D of $N \times N$. The element in line i and column j of D is marked as d_{ij}.

$$d_{ij} = \begin{cases} 1 \ |u(i) - u(j)| < r \\ 0 \ |u(i) - u(j)| \geq r \end{cases} i = 1 \sim N, j = 1 \sim N, i \neq j \tag{5}$$

Using the elements in D, the and $C_i^2(r)C_i^3(r)$.

$$C_i^2(r) = \sum_{j=1}^{N-1} d_{ij} \cap d_{(i+1)(\ j+1)} \tag{6}$$

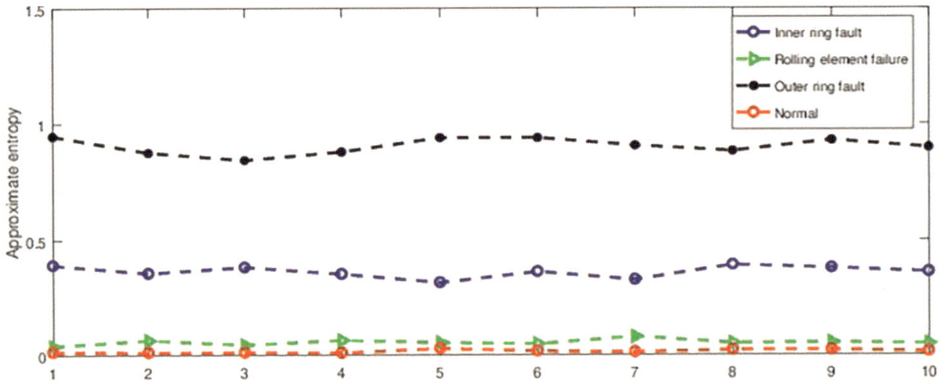

Figure 17.
Approximates the entropy curve.

$$C_i^3(r) = \sum_{j=1}^{N-2} d_{ij} \cap d_{(i+1)(j+1)} \cap d_{(i+1)(j+2)} \tag{7}$$

Calculate $H^2(r)$ and$H^3(r)$ from $C_i^2(r)$ and $C_i^3(r)$, and finally, calculate approximate entropy.

$$\text{ApEn}(\text{m, r, N}) = H^n(r) - H^{n+1}(r) \tag{8}$$

The approximate entropy of the sequence can be calculated from the above calculations.

3.2.3 Application of approximate entropy in mechanical fault diagnosis

Here, take six hundred vibration signals as a group and calculate ten groups of approximate entropy.

From **Figure 17**, it is not difficult to see that when the rolling bearing is normal, the value of the approximate entropy is not large, because, under normal circumstances, the generated signal is relatively single. When the rolling bearing fails, a lot of complicated information is generated, increasing the approximate entropy value. However, under the rolling body failure and normal working conditions, the two approximate entropy value is very similar, can not easily distinguish the two. If only a single entropy feature is used, it is easy to misjudge.

3.3 Sample entropy

3.3.1 The notion of sample entropy

In 2000, the concept of sample entropy was first proposed by Richman et al., a similar but more robust time-series complexity metric to approximate entropy, with greater resistance to interference and noise compared to approximate entropy [17].

Sample entropy improves the algorithm of approximate entropy and can reduce the error of approximate entropy when calculating. It is an algorithm similar to the approximate entropy but with superior computational accuracy.

Sample entropy has a better agreement. That is, if a time series has higher values than another time series, it also has higher values for other m and r values.

3.3.2 Algorithm for sample entropy

Generally, r takes 0.1~0.25SD (SD is the standard deviation of raw data), this paper$r = 0.15SD$, When m = 2 is selected, N = 500–1000.

Assuming the data is $\{X_i\} = \{x_1, x_2, \dots, x_N\}$, with its length N, a m-dimensional vector is reconstructed from the original signal:

$$x_i = [x_i, x_{i+1}, \dots, x_{i+m-1}], \mathrm{i} = 1, 2, \dots, \mathrm{N} - \mathrm{m} \tag{9}$$

Defines the distance between x_i and x_j:

$$d_{ij}d_{ij} = d[x(i), x(j)] = \max_{k \in [0, m-1]} [|x(i+k) - x(j+k)|] i, j = 1, 2, \dots, N - m, i \neq j \tag{10}$$

Then find the average value of $B_i^m(r)$

$$B^m(r) = \frac{1}{N - m} \sum_{i=1}^{N-m} B_i^m(r) \tag{11}$$

Similarly, again for the dimension m + 1, repeat the above steps, to obtain, and further obtain the final definition of the sample entropy, when N is a finite number

Similarly, repeat the above steps for dimension m+1 to obtain $B_i^{m+1}(r)$, and further obtain the final definition of sample entropy of $B^{m+1}(r)$. When n is finite,

$$SampEn(m, r, N) = \ln B^m(r) - \ln B^{m+1}(r) \tag{12}$$

The sample entropy of the sequence data is obtained from the above calculations.

3.3.3 Application of sample entropy in mechanical fault diagnosis

The first 6,000 data were taken in 600 sets to calculate the sample entropy, as shown in **Figure 18**.

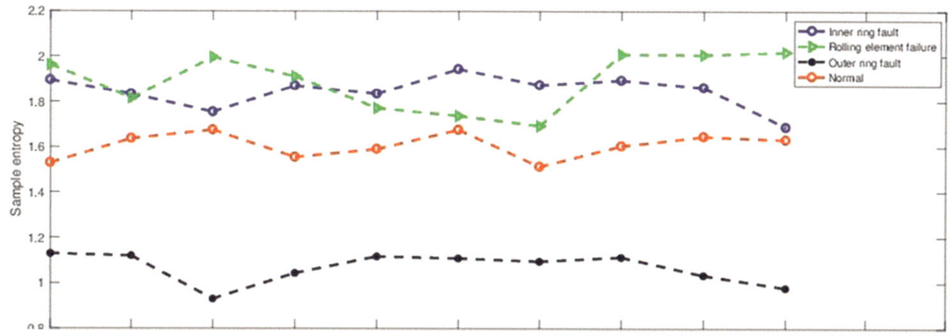

Figure 18.
Sample entropy curves.

From **Figure 18**, it is not difficult to see that the inner circle fault, the roll body fault, and the normal working conditions are very difficult to distinguish, but the outer circle fault can be distinguished.

3.4 Information entropy

3.4.1 The concept of information entropy

Information entropy is mostly used as a quantitative indicator of the information content of a system. The information entropy can be further used as a criterion for the optimization of the system equations [18].

3.4.2 Algorithm for information entropy

$$H(X) = -\sum p(xi) \log (p(xi)) \ (i = 1, 2, \dots, n) \tag{13}$$

X represents the random variable$(x1, x2, \dots xn)$, the value of which is, and p (xi) indicates the probability of an event xi.

3.4.3 Application of information entropy in mechanical fault diagnosis

The first 6000 data were taken in 600 sets and the information entropy was calculated in **Figure 19**.

As we can see from **Figure 19** the rolling body fault and the normal working condition intersect, and there are several very close places between the inner circle fault and the outer circle fault. If the information entropy is not processed, it is difficult to distinguish the rolling body fault and the normal working condition, and it is also easy to misjudge the inner circle fault and the outer ring fault.

3.5 Approximate entropy, sample entropy, information entropy, and maximum Lyapunov exponent

Taking 6000 data from each of four working conditions randomly, and taking 600 data as a group, the approximate entropy, sample entropy, information entropy, and maximum Lyapunov exponent are calculated. See **Tables 1–4** for specific results.

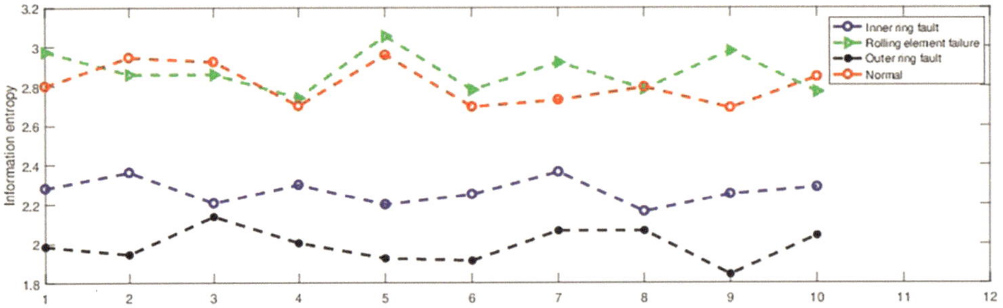

Figure 19.
Information entropy curve.

	Approximate entropy	Sample entropy	Information entropy	Maximum Lyapunov exponent
1 group	0.0145	1.5295	2.7995	0.8572
2 group	0.0122	1.6364	2.944	1.1640
3 group	0.0122	1.6746	2.9223	1.0358
4 group	0.0099	1.5549	2.6987	0.8828
5 group	0.0284	1.5904	2.9571	1.0570
6 group	0.0168	1.6752	2.6942	0.8198
7 group	0.0122	1.5137	2.7286	0.8005
8 group	0.0214	1.6044	2.7931	0.9247
9 group	0.0214	1.6455	2.6883	0.4819
10 group	0.0145	1.6316	2.8451	0.9472

Table 1.
Normal working condition.

	Approximate entropy	Sample entropy	Information entropy	Maximum Lyapunov exponent
1 group	0.3906	1.8951	2.2813	1.1505
2 group	0.3550	1.8330	2.3621	1.1941
3 group	0.3812	1.7537	2.2053	1.1570
4 group	0.3512	1.8701	2.2987	0.9440
5 group	0.3144	1.8343	2.1981	0.8531
6 group	0.3608	1.9440	2.2503	0.9038
7 group	0.3259	1.8736	2.3636	0.9706
8 group	0.3903	1.8948	2.1622	0.8296
9 group	0.3767	1.8621	2.2511	0.9782
10 group	0.3590	1.6874	2.2865	0.7633

Table 2.
Inner ring fault.

It can be seen from the above data that there is a certain correlation between the maximum Lyapunov exponent and entropy, but it can only show that the rolling bearings are in different degrees of chaotic systems under various working conditions. The maximum Lyapunov exponent has little discrimination in various working conditions, so it is still difficult to diagnose the fault of rolling bearings.

At the same time, we can see that the entropy with the highest correlation with the maximum Lyapunov exponent is also different under different working conditions of rolling bearings. For example, when the outer ring fails, the sample entropy can better improve the chaos phenomenon, but it is quite different from the maximum Lyapunov exponent under normal conditions.

In order to improve the chaotic phenomenon and diagnose the fault of rolling bearing, this paper does not use single entropy as the feature but uses approximate

	Approximate entropy	Sample entropy	Information entropy	Maximum Lyapunov exponent
1 group	0.0400	1.9635	2.9744	1.3315
2 group	0.0649	1.8153	2.8584	1.2093
3 group	0.0449	1.9966	2.8589	1.5054
4 group	0.0644	1.9115	2.7397	1.3533
5 group	0.0547	1.7711	3.0546	1.4797
6 group	0.0487	1.7368	2.7809	1.2739
7 group	0.0796	1.6927	2.9196	1.8167
8 group	0.0519	2.0092	2.7850	1.4544
9 group	0.0547	2.0074	2.9782	1.2152
10 group	0.0492	2.0193	2.7702	1.3233

Table 3.
Rolling element fault.

	Approximate entropy	Sample entropy	Information entropy	Maximum Lyapunov exponent
1 group	0.9484	1.1286	1.9842	1.0986
2 group	0.8777	1.1193	1.9447	0.8310
3 group	0.8442	0.9298	2.1366	0.5168
4 group	0.8804	1.0431	2.0030	0.8996
5 group	0.9411	1.1166	1.9253	0.9002
6 group	0.9405	1.1085	1.9136	0.6443
7 group	0.9071	1.0950	2.0648	0.7009
8 group	0.8832	1.1136	2.0648	0.5603
9 group	0.9296	1.0335	1.8442	0.9588
10 group	0.8982	0.9782	2.0407	0.9048

Table 4.
Outer ring fault.

entropy, sample entropy, and information entropy as the fault feature to diagnose the fault of rolling bearing.

3.6 Joint analysis

3.6.1 Feature extraction

The three entropy have considerable disadvantages under independent judgment, and it is easy to appear misjudgment, which needs to be further handled. So I began to extract the fault features of rolling bearings by jointly analyzing the approximate entropy, sample entropy, and information entropy.

The data are taken from the Western Reserve University, with a vibration acceleration signal in four different modes under a load of 2 horsepower, a fault diameter of 0.1778 mm, and a rotational speed of 1750 r/min. A total of four groups correspond to different patterns, with 60000 data, 10 segments, 6000 data per segment, with 10 segments, and 600 data per segment.

Each section finds an approximate entropy, sample entropy, information entropy, ten entropy values for each group, you can obtain ten approximate entropy mean, sample entropy mean, and information entropy mean. See **Figures 20–22**.

From the above figure, it is not difficult to see that some working conditions are still difficult to distinguish if you want to pass a single entropy feature. Approximate entropy has certain errors in the diagnosis of normal working condition and outer working condition, sample entropy has certain errors in the diagnosis of inner ring fault and rolling element fault, and information entropy has certain errors in the diagnosis of normal working condition and rolling element fault.

Therefore, as long as the three entropy means are extracted in each fault mode, four sets of column vectors are formed, such as **Table 5**, and each column corresponds to the entropy feature vector under the working condition.

The same entropy feature column vectors obtained from the same test data form the entropy feature matrix of the test data.

Based on the average entropy feature vector, we compare it with the test data entropy feature vector, that is, we take the absolute value after making the difference to obtain four new vectors.

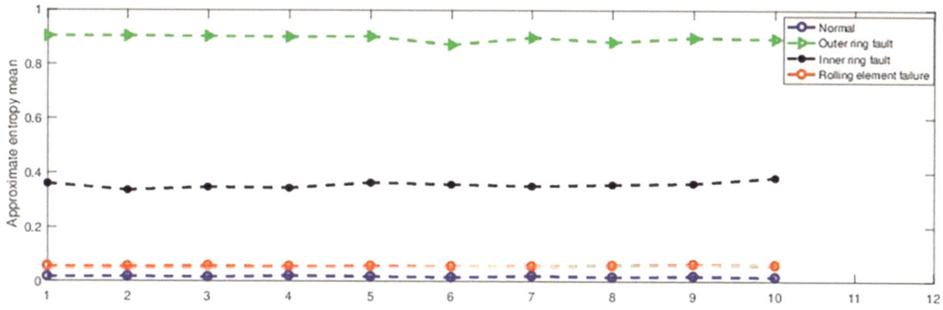

Figure 20.
Approximates the entropy mean.

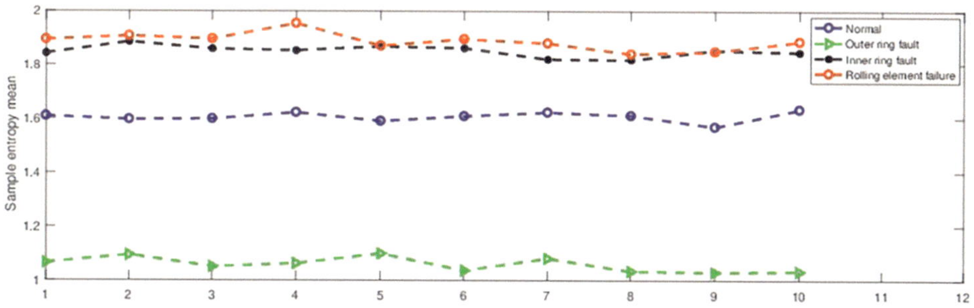

Figure 21.
Mean sample entropy.

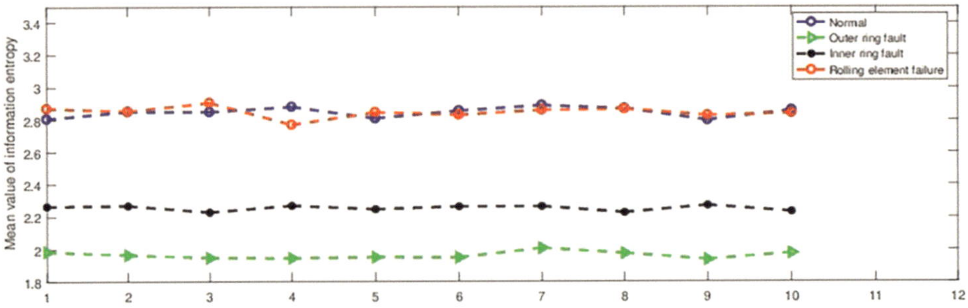

Figure 22.
Information entropy mean.

	Normal working condition	Outer ring fault	Inner ring fault	Ball fault
Approximate entropy mean	0.0164	0.9050	0.3605	0.0553
Sample entropy mean	1.6056	1.0666	1.8448	1.8923
Information entropy	2.8071	1.9922	2.2659	2.8720

Table 5.
Entropy feature matrices.

The four new vectors are formed into an entropy feature matrix, and the minimum of each row in the matrix is taken. The largest number of columns corresponding to the minimum is the judged test data closest to the failure mode.

However, in rare cases, three entropy features will determine the three failure modes. At this time, the failure mode corresponding to the approximate entropy mean with the greatest discrimination will be taken as the final failure mode of the test data.

3.6.2 Simulation experiment

After the early extraction of the rolling bearing fault characteristics, we have extracted the approximate entropy mean, sample entropy mean and information entropy mean in the four-fault modes of the rolling bearing. We can effectively distinguish the four-fault modes by comparing the feature vectors.

A matrix with a data quantity of 6000 was randomly generated using the data, and the fault features were extracted and determined to be a working condition. Later, the results were compared with the previous established standard, each group was tested 500 times, at least 10 groups, the test accuracy rate was based on the above description, and the simulation experiment began.

Also taken from Western Reserve University, the vibration acceleration signal in four different modes under a load of 2 HP, fault diameter of 0.1778 mm, and rotational speed of 1750 r/min.

The data of 6000 outer circle faults from the last 60000 data were divided into ten sections and 600 data for each section to calculate approximate entropy, sample entropy and fuzzy entropy mean to form a column vector.

The specific process is shown in **Figure 23**.

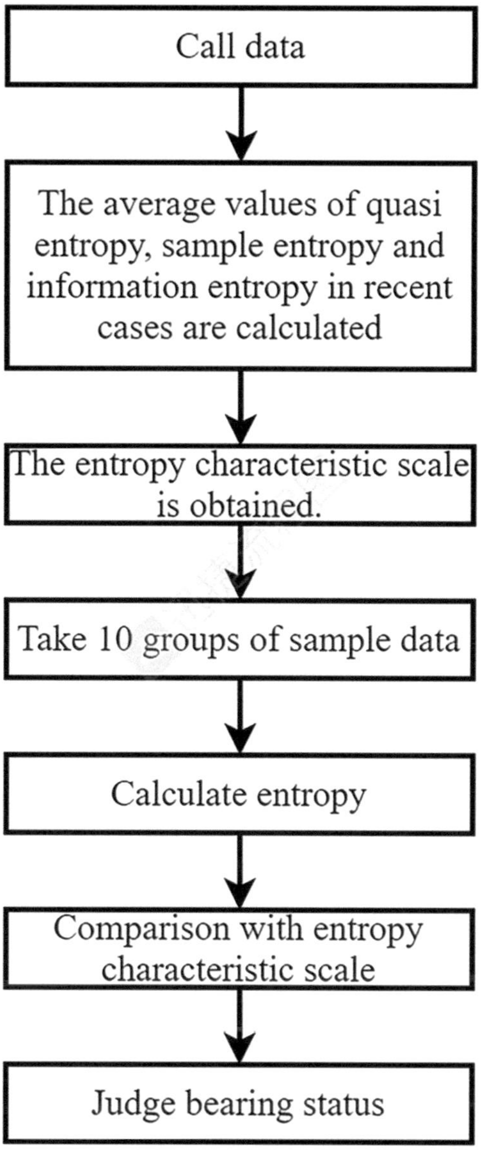

Figure 23.
Test process.

Choose any one of the four random working conditions as the test condition, extract 6000 vibration signals, and calculate the approximate entropy, sample entropy, and information entropy in groups of 600, and calculate the average value. The entropy mean is arranged as a test vector. The test vector is different from the corresponding components in the four working condition feature vectors, and the absolute value is taken. Then the test condition is the same as the condition corresponding to the feature vector with the smallest absolute value difference of each component. For the convenience of understanding and observation, the difference between the test vector and the feature vector group is visualized, resulting in **Figure 24**.

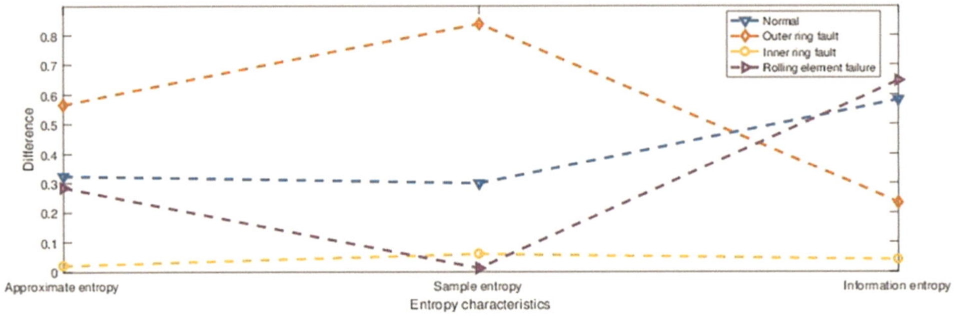

Figure 24.
Entropy difference 1.

It can be clearly seen from the figure that the absolute value of the difference between the yellow line characteristic vector corresponding to the inner ring fault and the test vector is the smallest, so the test condition is the inner ring fault (**Figure 25**).

It can be clearly seen from **Figure 16** that the absolute value of the difference between the purple line feature vector corresponding to the rolling element fault and the test vector is the smallest, so the test condition is the rolling element fault.

3.6.3 Comparative Experiment

In order to further prove the effectiveness of this method, a comparative experiment with the single entropy feature was carried out. The accuracy of the method using triple entropy combined features is 98.28%. The accuracy of the method characterized by single approximate entropy is 89.64%. The accuracy of the method characterized by single sample entropy is 79.80%. The accuracy of the method characterized by single information entropy is 71.96%. The four methods were repeated ten times respectively. **Table 6** shows the accuracy and average results of the experiment.

The accuracy of the triple entropy combination feature method is 8.64% higher than that of the single approximate entropy method, 18.48% higher than that of the single sample entropy method, and 98.28% higher than that of the single information entropy method. It can be seen that the method of triple entropy combination features can effectively diagnose the rolling bearing faults.

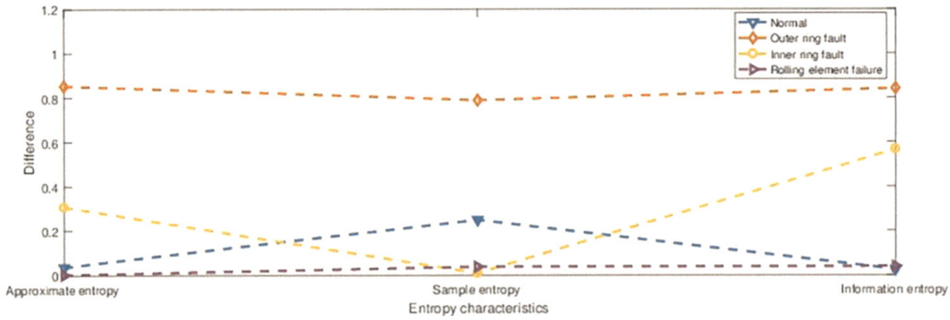

Figure 25.
Entropy difference 2.

	Approximate entropy	Sample entropy	Information entropy	Present method
1	90.80%	80.00%	71.60%	99.00%
2	89.60%	80.60%	75.80%	97.40%
3	91.00%	79.20%	70.00%	98.80%
4	86.80%	77.80%	74.20%	99.00%
5	89.60%	81.40%	71.20%	97.80%
6	91.20%	79.80%	70.60%	98.60%
7	90.00%	79.00%	72.40%	98.80%
8	89.40%	76.20%	69.00%	98.00%
9	89.20%	82.40%	71.60%	97.60%
10	88.80%	81.60%	71.20%	97.80%
Mean value	89.64%	79.80%	71.76%	98.28%

Table 6.
Comparison of accuracy of four methods.

3.7 Test results

Through many experiments, the accuracy rate of this method is above 97%, which is higher than other methods of the same type. The joint analysis of rolling bearing fault feature extraction based on approximate entropy, sample entropy, and information entropy is effective.

4. Conclusion

The chaotic system of rolling bearing is discussed, and the vibration signal of rolling bearing is changed by fast Fourier. It is found that different chaotic phenomena appear in different working conditions. Therefore, a fault diagnosis method of rolling bearings based on entropy features is proposed, and the relationship between approximate entropy, sample entropy, information entropy, and the maximum Lyapunov exponent is studied. It is concluded that single entropy can't improve the chaotic phenomenon. Furthermore, a fault extraction method of rolling bearing combining approximate entropy, characteristic entropy, and information entropy is proposed. The feasibility of this method is proved by the comparative experiment of single entropy and approximate entropy sample entropy information entropy joint analysis. In the final test, the classification accuracy reached 98.28%. The feasibility of this method is proved, and four working conditions of rolling bearings are diagnosed with high precision.

Nomenclature

c	Radius of rolling bearing inner ring
r_2	Radius of rolling bearing outer ring
D	Rolling element diameter of bearing

D	bearing joint diameter
Z	Number of rolling bodies
Θ	bearing contact angle
f_r	Rotation frequency of rolling bearing
f_i	Inner ring fault frequency
f_o	Outer ring fault frequency
f_{BS}	Ball failure frequency
E	Base of natural logarithm
I	Imaginary unit

Author details

Yongjian Sun* and Zihan Wang
University of Jinan, Jinan, Shandong, China

*Address all correspondence to: sunyongjian2006@163.com

References

[1] Li X, Zhang W, Ding Q. Understanding and improving deep learning-based rolling bearing fault diagnosis with attention mechanism. Signal Processing. 2019;**161**:136-154

[2] Cao H, Niu L, Xi S, et al. Mechanical model development of rolling bearing-rotor systems: A review. Mechanical Systems and Signal Processing. 2018;**102**: 37-58

[3] Li X, Zhang W, Ding Q, et al. Multi-layer domain adaptation method for rolling bearing fault diagnosis. Signal Processing. 2019;**157**:180-197

[4] Wang Z, Yao L, Cai Y. Rolling bearing fault diagnosis using generalized refined composite multiscale sample entropy and optimized support vector machine. Measurement. 2020;**156**: 107574

[5] Li X, Jiang H, Niu M, et al. An enhanced selective ensemble deep learning method for rolling bearing fault diagnosis with beetle antennae search algorithm. Mechanical Systems and Signal Processing. 2020; **142**:106752

[6] Meng Z, Zhan X, Li J, et al. An enhancement denoising autoencoder for rolling bearing fault diagnosis. Measurement. 2018;**130**:448-454

[7] Xu Y, Zhang K, Ma C, et al. Adaptive Kurtogram and its applications in rolling bearing fault diagnosis. Mechanical Systems and Signal Processing. 2019;**130**: 87-107

[8] Cerrada M, Sánchez RV, Li C, et al. A review on data-driven fault severity assessment in rolling bearings. Mechanical Systems and Signal Processing. 2018;**99**:169-196

[9] Wang Z, Yao L, Cai Y, et al. Mahalanobis semi-supervised mapping and beetle antennae search based support vector machine for wind turbine rolling bearings fault diagnosis. Renewable Energy. 2020;**155**:1312-1327

[10] Li X, Jiang H, Wang R, et al. Rolling bearing fault diagnosis using optimal ensemble deep transfer network. Knowledge-Based Systems. 2021;**213**: 106695

[11] Chen X, Yang Z, Lou W. Fault diagnosis of rolling bearing based on the permutation entropy of VMD and decision tree. In: 2019 3rd International Conference on Electronic Information Technology and Computer Engineering (EITCE). 2019. pp. 1911-1915

[12] Jiao J, Zhao M, Lin J, et al. Hierarchical discriminating sparse coding for weak fault feature extraction of rolling bearings. Reliability Engineering & System Safety. 2019;**184**: 41-54

[13] Akinlar MA, Tchier F, Inc M. Chaos control and solutions of fractional-order Malkus waterwheel model. Chaos, Solitons & Fractals. 2020;**135**:109746

[14] Li X, Song S, Wu J. Exponential stability of nonlinear systems with delayed impulses and applications. IEEE Transactions on Automatic Control. 2019;**64**(10):4024-4034

[15] Strogatz SH. Nonlinear Dynamics and Chaos: With Applications to Physics, Biology, Chemistry, and Engineering. CRC Press; 2018

[16] Wang Z, Yao L, Cai Y. Rolling bearing fault diagnosis using generalized refined composite multiscale sample entropy and optimized support vector

machine. Measurement. 2020;**156**: 107574

[17] Zhao D, Liu S, Cheng S, et al. Parallel multi-scale entropy and it's application in rolling bearing fault diagnosis. Measurement. 2021;**168**:108333

[18] Yan X, Liu Y, Huang D, et al. A new approach to health condition identification of rolling bearing using hierarchical dispersion entropy and improved Laplacian score. Structural Health Monitoring. 2021;**20**(3): 1169-1195

Chapter 5

Perspective Chapter: Dynamic Analysis of High-Rise Buildings Using Simplified Numerical Method

Emarti Kumari

Abstract

This chapter emphasizes on the static and dynamic characteristics of multi-story building subjected to uniformly distributed and wind load. First-order shear deformation theory is used to formulate governing equations based on the finite element method. The multi-story building is considered as a vertical cantilever beam/plate and modeled using nine-node degenerated shell element. Fictitious membrane and shear stresses are eliminated by considering Mixed Interpolation Tonsorial Component (MITC) technique. Here, the static and dynamic characteristics of multi-story buildings have been investigated take into account as a vertical cantilever plate subjected to UDL, triangular load (wind load) and combination of both. In this chapter authors demonstrated the deformation shapes, longitudinal stress and in-plane shear stress and principle strains in various loading conditions of vertical cantilever flat panel. Moreover, investigated the dynamic characteristics of multi-story buildings considering as a vertical cantilever plates and governing equations of motion are derived by employing Hamilton's principle. Moreover, nonlinear transient response of high-rise structures has been studied here by employing the energy and momentum conservation implicit time integration scheme. The structural analysis of tall buildings has been carried out here through commercial software ANSYS. Matrix amplitude method is employed to investigate the large-amplitude flexural vibration responses of flat panels. Also, plotted the fast Fourier transform and phase portraits for first three bending modes.

Keywords: static, dynamic, high rise building, wind load, displacements, critical buckling load

1. Introduction

In the recent times, dynamically increase in population and limitation of land acquisition high-rise structures or multi-story building become major concern of researchers. The critical issue in high-rise buildings is selecting proper structural form to resist the lateral and flexural loads. Therefore, the static and dynamic responses of high-rise building is essential to design safe and reliable structure for engineers. These high-rise buildings may be transformed into multi-story buildings in order to make more floors space but occupy less lands space [1]. These multistory buildings are used for residential flats, offices, shopping complex, Malls, Hotels and public centers.

Lateral loads such as wind load and seismic load play predominant role in high-rise structures.

1.1 Wind loading

There are several instances where structures have failed owing to an instability that needs second-order analysis (P-Delta). One of the problems resulted from wind loading. The wind induces outward and inward as a triangular load act on the surfaces of multi-story buildings. Structural instability issues arises when structures could not bear certain loads and buckling of structures +occurs due to dead load, wind load and seismic load like earthquake [2]. Ankireddi and Yang [3] considered gradually increased load along building height (wind load) as shown in **Figure 1**. To simplify the analysis [4, 5] load is reduced by 50 percent at the center of building as consideration of BS EN 1991-1-4:2005.

The direction of wind is very important throughout the life of structure. Wen [6] found that the wind direction analysis is not conducted seriously. Analysis of wind direction becomes necessary for the case of high-rise buildings and suspension bridges. Moreover, wind speed is also plays crucial role on construction sites, structures will collapse without warning due to speed of wind. For example, building, power lines and trees are collapsed in Hurricane Frederic east of Pascagoula, Mississippi on September 12, 1979 in coastal areas. Mehta et al. [7] estimated the effect of wind speed based on an indirect approach. Authors collected comprehensive data to introduce an alternate approach and convenient equipment. It is noticed that wind speed and direction significantly contribute in deflection of multi-story structures.

1.2 Urban context and sustainability

For energy saving and sustainable development high-rise building requires sunlight access, considering optimum utilization of limited conventional resources.

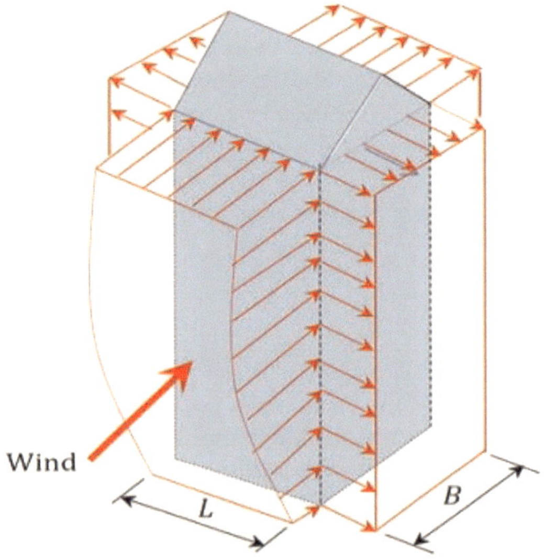

Figure 1.
Wind loads act on the surface of the building [3, 5].

Therefore, new multi-story buildings are optimized on its shape, height and orientation considering environmental factors such as harmony between buildings and their urban and environmental contexts.

Environment friendly structures are classified into three basic sets of strategies. First one is the minimization of operation cost and material consumption; second one minimization of energy consumption or maximum utilization of solar energy and sun light; third one is consideration of whether conditions according to different climate zones.

To design building for different climatic zones, then consider the hot seasons and the cold seasons to deal with, and you have the two mid- seasons. So, you have to design the enclosure, the skin as a responsive environmental filter for energy efficiency.

For the analysis purpose, a cantilever beam (l = 60 m, b = 12 m and h = 1 m) with fixed support at bottom and free at remaining all three ends is modeled as high-rise building. The analysis is performed considering the building experiencing different loading conditions viz., uniformly distributed load, uniformly varying load (wind load) with zero magnitude at bottom and the combination of the two said loads.

The static behavior of hybrid high rise buildings have been studied here considering a vertical cantilever flat panel under uniformly distributed load, wind load and combination of both. Linear bending response of vertical cantilever plates has been analyzed under triangular load. The structural analysis of tall buildings has been carried out here through commercial software ANSYS.

1.3 High-rise building considered as cantilever beams

An approximate method for the static and dynamic analysis of high-rise buildings is continuum method in which these structures are substituted by a continuum beam/plate, adopting Euler-Bernoulli or Timoshenko beam theory/Classical or Mindlin or higher-order plate theory. The bending analysis of tall building have been performed by employing continuum-based Kwan model, here, tall buildings were considered as a cantilever beam and studied bending behavior of tall building by employing Euler-Bernoulli beam theory [8] and also compared the analytical results with numerical results obtained by ETABS software. Framed tube structures primarily act like cantilever hollow box beams. These beams could resist more moments of the lateral loads; the beam bending action of the framed tube structures were complicated due to shear lag in the web and flange panel; [9] introduced a simple hand-calculation method considering shear lag effect for approximate static analysis of framed tube structures. Alavi et al. [10, 11] proposed a simple mathematical technique to design minimum cost tall and slender structures that might be used as conceptual/early-stage design. Authors considered the tall structures as cantilever beam and studied the flexural vibration behavior [10] and peak lateral deflection response [11] of high-rise buildings. They also performed the parametric study considering 42-story and 60-story buildings.

Most of the researchers modeled the tall building as cantilever beam to study the free vibration behavior [12–16]. The discrepancy of empirical formulas given by various researchers for the fundamental frequency of tall buildings was examined by Dym et al. [12] and used the Euler-Bernoulli and Timoshenko beam model to estimate the natural frequencies of these structures. The natural frequency of multistory building under seismic load was determined by Kaviani et al. [14] using

Timoshenko beam theory. They assumed each lateral load carrying subcomponent (floor) of tall building is considered continuous cantilever beam with variable cross-section.

The natural frequencies of framed tube and shear walls structures have been calculated by using energy method and Hamilton's principle [17]. Authors considered the framed tube and shear walls structures as a cantilever beam based on continuum approach. They also compared the analytical results with numerical results obtained by SAP2000. Piccardo et al. [18] introduced an equivalent three-dimensional space continuous Timoshenko beam model to study the static and dynamic behavior of tower building.

Hallebrand and Jakobsson [19] investigated the high-rise building under static and dynamic loads to investigate the effect of deflections, resonance frequencies, accelerations and stability. They discussed the modeling techniques and issues to model high-rise building using finite element method. Authors also compared the vertical loading (such as self-weight, imposed loads, snow loads and live loads) and horizontal loading (for example wind load or design load and unintended inclinations) considering different modeling techniques.

Miranda [20] introduced an approximate method based on an equivalent continuum beam model to determine lateral displacement and maximum inter-story drift of tall buildings under seismic load. Multistory structures were considered as an equivalent continuum model taking in account the combination of shear cantilever beam and flexural cantilever beam. Author examined the effect of lateral force along the height of tall structures, shear deformations, flexural deformations on the multistory buildings that was considered as a beam structure. An approximate method was further generalized [21] to consider the non-uniform lateral stiffness of tall structures. The non-uniform Timoshenko beam was modeled [22] to describe the spectrum analysis of it. They assessed the first four vibration modes of non-uniform Timoshenko beam model under seismic loading conditions. Attention of researchers is also focused on the eigenvalue solutions of beams (Rafezy and Howson 2008) that would be considered as a high-rise building.

Authors considered the step change of properties along the height of the structures. Recently, the nonlinear static and dynamic response of thin-wall composite structures were investigated [23–26] by employing finite element method with first-order shear deformation theory. Here, author considered a high-rise building as a cantilever plat to study the static and dynamic behavior of these thin-wall structures.

2. Finite element formulation

The static and dynamic analysis of high-rise buildings under uniform pressure and wind load are performed by using finite element method based commercial software ANSYS. To discretize the high-rise building type cantilever beam/plate considered the eight-node 281 shell element. To solve the multi-dimensional problem using Green-Gauss theorem will expressed as:

$$\int_V \sigma^T \varepsilon(\delta) dV - \int_V \delta^T f dV - \int_S \begin{bmatrix} (n_x\sigma_x + n_y\tau_{xy} + n_z\tau_{xz})\delta_x \\ +(n_x\tau_{xy} + n_y\sigma_y + n_z\tau_{yz})\delta_y \\ +(n_x\tau_{xz} + n_y\tau_{yz} + n_z\sigma_z)\delta_z \end{bmatrix} dS \tag{1}$$

Here, σ represents six independent component of stress $\sigma = \left[\sigma_x, \sigma_y, \sigma_z, \tau_{yz}, \tau_{xz}, \tau_{xy}\right]^T$ normal stresses and shear stresses; ε represents six strains corresponding to stresses $\varepsilon = \left[\varepsilon_x, \varepsilon_y, \varepsilon_z, \gamma_{yz}, \gamma_{xz}, \gamma_{xy}\right]^T$; dV is volume integration $dV = dx\, dy\, dz$; distributed force per unit volume $f = \left[f_x, f_y, f_z\right]^T$; displacement vector is $\delta = \left[u_x, u_y, u_z, \theta_x, \theta_y, \theta_y\right]^T$; the unit normal to surface dA is $n = \left[n_x, n_y, n_z\right]^T$.

2.1 Constitutive equations

The kinematic correlations and the mechanical and thermodynamic concepts are applicable at all continuum irrespective of its physical constitutions. Here, considered the equations characterizing the individual material and its reaction to apply loads. These equations are known as constitutive eqs.

A material body said to be isotropic/homogeneous if the properties of material are same throughout the body. In an anisotropic/heterogeneous body, the properties of material are function of position.

A material body supposed to be ideally elastic under isothermal conditions, the body will recover its original form with removal of forces causing deformation. One-to-one relationship (based on generalized Hooke's law) between the state of stress and state of strain will be written as:

$$\begin{Bmatrix} \sigma_{11} \\ \sigma_{22} \\ \sigma_{33} \\ \sigma_{23} \\ \sigma_{13} \\ \sigma_{12} \end{Bmatrix} = \begin{bmatrix} C_{11} & C_{12} & C_{13} & C_{14} & C_{15} & C_{16} \\ C_{21} & C_{22} & C_{23} & C_{24} & C_{25} & C_{26} \\ C_{31} & C_{32} & C_{33} & C_{34} & C_{35} & C_{36} \\ C_{41} & C_{42} & C_{43} & C_{44} & C_{45} & C_{46} \\ C_{51} & C_{52} & C_{53} & C_{54} & C_{55} & C_{56} \\ C_{61} & C_{62} & C_{63} & C_{64} & C_{65} & C_{66} \end{bmatrix} \begin{Bmatrix} \varepsilon_{11} \\ \varepsilon_{22} \\ \varepsilon_{33} \\ \varepsilon_{23} \\ \varepsilon_{13} \\ \varepsilon_{12} \end{Bmatrix} \tag{2}$$

Where C_{ij} is the elastic coefficient.

The elastic coefficient matrix C_{ij} is a symmetric ($C_{ij} = C_{ji}$) therefore, there is 21 independent coefficients of the matric [C].

For three perpendicular planes (x-y, x-z and y-z) to each other known orthogonal planes due to symmetry the number of elastic coefficients are reduced to nine, and these materials are known as orthotropic. The stress-strain relations for an orthotropic material will be expressed as:

$$\begin{Bmatrix} \sigma_{11} \\ \sigma_{22} \\ \sigma_{33} \\ \sigma_{23} \\ \sigma_{13} \\ \sigma_{12} \end{Bmatrix} = \begin{bmatrix} C_{11} & C_{12} & C_{13} & 0 & 0 & 0 \\ C_{21} & C_{22} & C_{23} & 0 & 0 & 0 \\ C_{31} & C_{32} & C_{33} & 0 & 0 & 0 \\ 0 & 0 & 0 & C_{44} & 0 & 0 \\ 0 & 0 & 0 & 0 & C_{55} & 0 \\ 0 & 0 & 0 & 0 & 0 & C_{66} \end{bmatrix} \begin{Bmatrix} \varepsilon_{11} \\ \varepsilon_{22} \\ \varepsilon_{33} \\ \varepsilon_{23} \\ \varepsilon_{13} \\ \varepsilon_{12} \end{Bmatrix} \tag{3}$$

The inverse relations, strain–stress relations may be written as:

$$\begin{Bmatrix} \varepsilon_{11} \\ \varepsilon_{22} \\ \varepsilon_{33} \\ \varepsilon_{23} \\ \varepsilon_{13} \\ \varepsilon_{12} \end{Bmatrix} = \begin{bmatrix} S_{11} & S_{12} & S_{13} & 0 & 0 & 0 \\ S_{21} & S_{22} & S_{23} & 0 & 0 & 0 \\ S_{31} & S_{32} & S_{33} & 0 & 0 & 0 \\ 0 & 0 & 0 & S_{44} & 0 & 0 \\ 0 & 0 & 0 & 0 & S_{55} & 0 \\ 0 & 0 & 0 & 0 & 0 & S_{66} \end{bmatrix} \begin{Bmatrix} \sigma_{11} \\ \sigma_{22} \\ \sigma_{33} \\ \sigma_{23} \\ \sigma_{13} \\ \sigma_{12} \end{Bmatrix} \tag{4}$$

$$\begin{Bmatrix} \varepsilon_{11} \\ \varepsilon_{22} \\ \varepsilon_{33} \\ \varepsilon_{23} \\ \varepsilon_{13} \\ \varepsilon_{12} \end{Bmatrix} = \begin{bmatrix} \frac{1}{E_1} & -\frac{\nu_{21}}{E_2} & -\frac{\nu_{31}}{E_3} & 0 & 0 & 0 \\ -\frac{\nu_{12}}{E_1} & \frac{1}{E_2} & -\frac{\nu_{32}}{E_3} & 0 & 0 & 0 \\ -\frac{\nu_{13}}{E_1} & -\frac{\nu_{23}}{E_2} & \frac{1}{E_3} & 0 & 0 & 0 \\ 0 & 0 & 0 & \frac{1}{G_{23}} & 0 & 0 \\ 0 & 0 & 0 & 0 & \frac{1}{G_{13}} & 0 \\ 0 & 0 & 0 & 0 & 0 & \frac{1}{G_{12}} \end{bmatrix} \begin{Bmatrix} \sigma_{11} \\ \sigma_{22} \\ \sigma_{33} \\ \sigma_{23} \\ \sigma_{13} \\ \sigma_{12} \end{Bmatrix} \tag{5}$$

Here, $[S]_{6\times 6}$ denotes the compliance coefficients; $[S] = [C]^{-1}$; E_1, E_2, E_3 are Young's modulus in 1 (longitudinal), 2 (transverse) and 3 (normal) material directions, respectively; Poisson's ratio ν_{ij} is the ratio of transverse strain in jth direction to the axial strain in ith direction when load is applied along longitudinal direction or stressed in ith direction; G_{12}, G_{13}, and G_{23} are shear moduli in the x-y, x-z and y-z planes, respectively. The compliance matrix [S] is symmetric matrix, because compliance matrix is the inverse of stiffness matrix. Symmetric matrix inverse is also symmetric.

Therefore: $\frac{\nu_{21}}{E_2} = \frac{\nu_{12}}{E_1}$; $\frac{\nu_{31}}{E_3} = \frac{\nu_{13}}{E_1}$; $\frac{\nu_{32}}{E_3} = \frac{\nu_{23}}{E_2}$ or $\frac{\nu_{ij}}{E_i} = \frac{\nu_{ji}}{E_j}$ (no sum on i, j).

Here i, j = 1,2,3. Hence, there are only nine independent material coefficients (E_1, E_2, E_3, G_{12}, G_{13}, G_{23}, ν_{12}, ν_{13}, ν_{23}) for an orthotropic material. For an isotropic material (material having infinite number of planes of material symmetry) independent elastic coefficients are reduced to two ($E_1 = E_2 = E_3 = E$, $\nu_{12} = \nu_{13} = \nu_{23} = \nu$, $G_{12} = G_{13} = G_{23} = G = E/2(1 + \nu)$.

The state of plane stress is expressed to be one in which transverse stresses are neglected. Then, for the orthotropic material the strain-stress relations to describe the state of plane stress is:

$$\begin{Bmatrix} \varepsilon_1 \\ \varepsilon_2 \\ \varepsilon_{12} = \varepsilon_6 \end{Bmatrix} = \begin{bmatrix} S_{11} & S_{12} & 0 \\ S_{12} & S_{22} & 0 \\ 0 & 0 & S_{66} \end{bmatrix} \begin{Bmatrix} \sigma_1 \\ \sigma_2 \\ \sigma_{12} = \sigma_6 \end{Bmatrix}$$
$$= \begin{bmatrix} 1/E_1 & -\nu_{21}/E_2 & 0 \\ -\nu_{12}/E_1 & 1/E_2 & 0 \\ 0 & 0 & 1/G_{12} \end{bmatrix} \begin{Bmatrix} \sigma_1 \\ \sigma_2 \\ \sigma_{12} = \sigma_6 \end{Bmatrix} \tag{6}$$

The strain-stress relation expressed in Eq. (6) are inverted to obtain the stress-strain relations:

$$\begin{Bmatrix} \sigma_1 \\ \sigma_2 \\ \sigma_{12} = \sigma_6 \end{Bmatrix} = \begin{bmatrix} Q_{11} & Q_{12} & 0 \\ Q_{12} & Q_{22} & 0 \\ 0 & 0 & Q_{66} \end{bmatrix} \begin{Bmatrix} \varepsilon_1 \\ \varepsilon_2 \\ \varepsilon_{12} = \varepsilon_6 \end{Bmatrix} \tag{7}$$

Here, Q_{ij} is known as plane stress-reduced stiffness, are expressed by:

$$\begin{aligned} Q_{11} &= \frac{S_{22}}{S_{11}S_{22} - S_{12}^2} = \frac{E_1}{1 - \nu_{12}\nu_{21}} \\ Q_{12} &= \frac{S_{12}}{S_{11}S_{22} - S_{12}^2} = \frac{\nu_{12}E_2}{1 - \nu_{12}\nu_{21}} \\ Q_{22} &= \frac{S_{11}}{S_{11}S_{22} - S_{12}^2} = \frac{E_2}{1 - \nu_{12}\nu_{21}} \\ Q_{66} &= \frac{1}{S_{66}} = G_{12} \end{aligned} \tag{8}$$

Thus, reduced stiffness involved four independent material constants E_1, E_2, ν_{12}, G_{12}.

The transverse shear stresses and shear strain relations for orthotropic materials are defined as:

$$\begin{Bmatrix} \sigma_{23} = \sigma_4 \\ \sigma_{13} = \sigma_5 \end{Bmatrix} = \begin{bmatrix} Q_{44} & 0 \\ 0 & Q_{55} \end{bmatrix} \begin{Bmatrix} \varepsilon_{23} = \varepsilon_4 \\ \varepsilon_{13} = \varepsilon_5 \end{Bmatrix} \tag{9}$$

Here, $Q_{44} = C_{44} = G_{23}$ and $Q_{55} = C_{55} = G_{13}$.

2.2 Transformation of components

In structural analysis, it is required to consider all the quantities for common structural coordinate system. Scalars are independent of any coordinate system, whereas vectors and tensors are independent of a particular coordinate system, and their components are not. The same vectors and tensors have different components in different coordinate systems, but any two sets of components of a vectors and tensor will be related by writing one set of components in terms of the other. Transformation of vectors component considering barred $(\overline{x}_1, \overline{x}_2, \overline{x}_3)$ and unbarred (x_1, x_2, x_3) coordinate systems are related as shown in **Figure 2** and written in Eqs. (10) and (11).

$$\begin{aligned} x_1 &= x_1(\overline{x}_1, \overline{x}_2, \overline{x}_3) \\ x_2 &= x_2(\overline{x}_1, \overline{x}_2, \overline{x}_3) \\ x_3 &= x_3(\overline{x}_1, \overline{x}_2, \overline{x}_3) \end{aligned} \tag{10}$$

Inverse relations are written as:

$$\begin{aligned} \overline{x}_1 &= \overline{x}_1(x_1, x_2, x_3) \\ \overline{x}_2 &= \overline{x}_2(x_1, x_2, x_3) \\ \overline{x}_3 &= \overline{x}_3(x_1, x_2, x_3) \end{aligned} \tag{11}$$

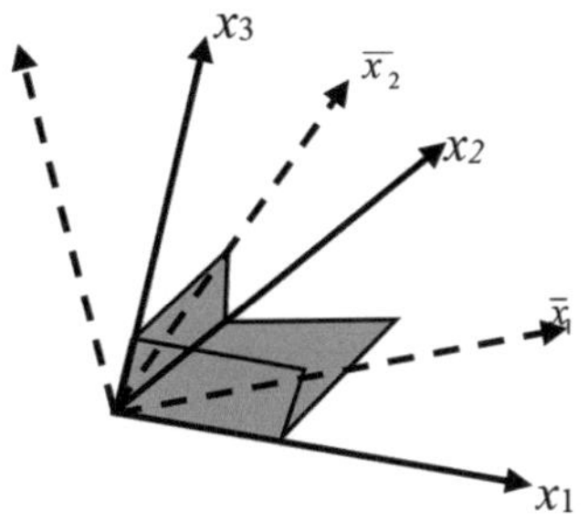

Figure 2.
Unbarred and barred rectangular coordinate system.

2.3 Transformation of material stiffness

The material stiffness C_{ijkl} is the fourth order tensor. Thus, considering the in general law of the fourth order tensor transforms as given in Eq. (12).

$$\overline{C}_{ijkl} = a_{im}a_{jn}a_{kp}a_{lq}C_{mnpq} \tag{12}$$

Schematic representation of rectangular plate with global and material coordinates is shown in **Figure 3**. For the plane stress case, the elastic stiffness Q_{ij} in the principal material system are related to $\overline{Q}_{ij}$ in the reference coordinate system is written as:

$$\begin{aligned}
\overline{Q}_{11} &= Q_{11}\cos^4\theta + 2(Q_{12}+2Q_{66})\sin^2\theta\cos^2\theta + Q_{22}\sin^4\theta \\
\overline{Q}_{12} &= (Q_{11}+Q_{22}-4Q_{66})\sin^2\theta\cos^2\theta + Q_{12}(\sin^4\theta + \cos^4\theta) \\
\overline{Q}_{22} &= Q_{11}\sin^4\theta + 2(Q_{12}+2Q_{66})\sin^2\theta\cos^2\theta + Q_{22}\cos^4\theta \\
\overline{Q}_{16} &= (Q_{11}-Q_{12}-2Q_{66})\sin\theta\cos^3\theta + (Q_{12}-Q_{22}+2Q_{66})\sin^3\theta\cos\theta \\
\overline{Q}_{26} &= (Q_{11}-Q_{12}-2Q_{66})\sin^3\theta\cos\theta + (Q_{12}-Q_{22}+2Q_{66})\sin\theta\cos^3\theta \\
\overline{Q}_{66} &= (Q_{11}+Q_{22}-2Q_{12}-2Q_{66})\sin^2\theta\cos^2\theta + Q_{66}(\sin^4\theta + \cos^4\theta) \\
\overline{Q}_{44} &= Q_{44}\cos^2\theta + Q_{55}\sin^2\theta \\
\overline{Q}_{45} &= (Q_{55}-Q_{44})\cos\theta\sin\theta \\
\overline{Q}_{55} &= Q_{55}\cos^2\theta + Q_{44}\sin^2\theta
\end{aligned} \tag{13}$$

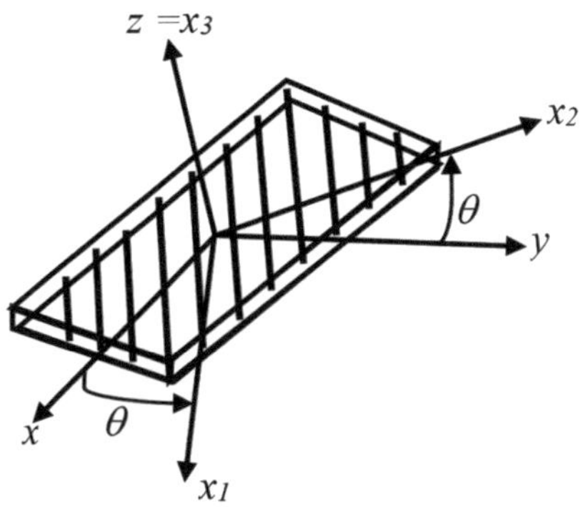

Figure 3.
Rectangular plate with global and material coordinate systems.

2.4 Governing equations

The total potential energy of the general elastic body is written as:

$$\prod = \frac{1}{2}\int_V \sigma^T \varepsilon dV - \int_V \delta^T f dV - \int_S \delta^T T dS - \sum_i \delta_i^T P_i \tag{14}$$

Here, σ = [D][B]{δ} and ε = [B]{δ}; [D] is the flexural rigidity matrix and [B] is the strain–displacement matrix.

2.5 Static analysis

Firstly, for the static analysis of multi-story building are performed by solving this governing equation that may be expressed as:

$$[K_L + K_{NL1}(\delta) + K_{NL2}(\delta,\ \delta)]\{\delta\} = \{F_P\} + \{F_{wind}\} \tag{15}$$

Here, $[K_L]$ is linear stiffness matrix; $[K_{NL}]$ is nonlinear stiffness matrix; $\{\delta\}$ is displacement vector represents six degree of freedom three displacements u_x, u_y, u_z and three rotations θ_x, θ_y, θ_z along x-axis, y-axis and z-axis, respectively; $\{F_p\}$ and $\{F_{wind}\}$ are uniformly distributed load and wind load vector.

2.6 Dynamic analysis

Thereafter, for the dynamic analysis of multi-degree of freedom systems the governing equation of motion may be written as:

$$[M]\{\ddot{\delta}\} + [K_L + K_{NL1}(\delta) + K_{NL2}(\delta,\ \delta)]\{\delta\} = \{F_p\} + \{F_{wind}\} \tag{16}$$

Here, $[M]$ is mass matrix; $\{\ddot{\delta}\}$ is acceleration vector.

For the dynamic forced vibration analysis of large story building Eq. (16) used here, whereas for the free vibration analysis of high-rise building governing equation may be written as:

$$[M]\{\ddot{\delta}\} + [K_L + K_{NL1}(\delta) + K_{NL2}(\delta,\ \delta)]\{\delta\} = 0 \tag{17}$$

The natural frequency and deformations of high-rise building are expressed by Eigen-values and Eigen-vector solutions of Eq. (17).

$$\{-[M]\omega^2 + [K_L + K_{NL1}(\delta) + K_{NL2}(\delta,\ \delta)]\}\{\delta\} = 0 \tag{18}$$

Here, ω is the natural frequency of high-rise building represented as an Eigen-values; $\{\delta\}$ is the deformation of structures along six degree of freedom displacements (u_x, u_y, u_z) and rotations (θ_x, θ_y, θ_z) is represented by Eigen-vector of Eq. (17).

2.7 Solution Procedure for large amplitude flexural vibration analysis

Assume a harmonic solution of the displacement vector $\delta = \delta_{\max} \sin \theta t$, the weighted residual of Eq. (16) with $\{F_P + F_{wind} = 0\}$ along the path t = 0 to $T/4$ (δ = 0 to δ_{max}) may be expressed as:

$$\int_0^{T/4} R \sin \theta t dt = \{0\} \tag{19}$$

Where, the residual of Eq. (19) is:

$$\{R\} = \left[K_L + K_{NL1}(\delta_{\max}) \sin \theta t + K_{NL2}(\delta_{\max}, \ \delta_{\max}) \sin^2 \theta t - \theta^2 M\right] \{\delta_{\max}\} \sin \theta t \tag{20}$$

Evaluating the integral of Eq. (20); the matrix amplitude equation may be written as [27]:

$$\left[K_L + \frac{4}{3\pi} K_{NL1}(\delta_{\max}) + \frac{3}{4} K_{NL2}(\delta_{\max}, \ \delta_{\max}) - \theta^2 M\right] \{\delta_{\max}\} = \{0\} \tag{21}$$

The matrix amplitude Eq. (21) is solved iteratively (Naghsh and Azhari 2015) to find the frequency verses amplitude relationship of cantilever plates.

2.8 Flow chart for static analysis through ANSYS

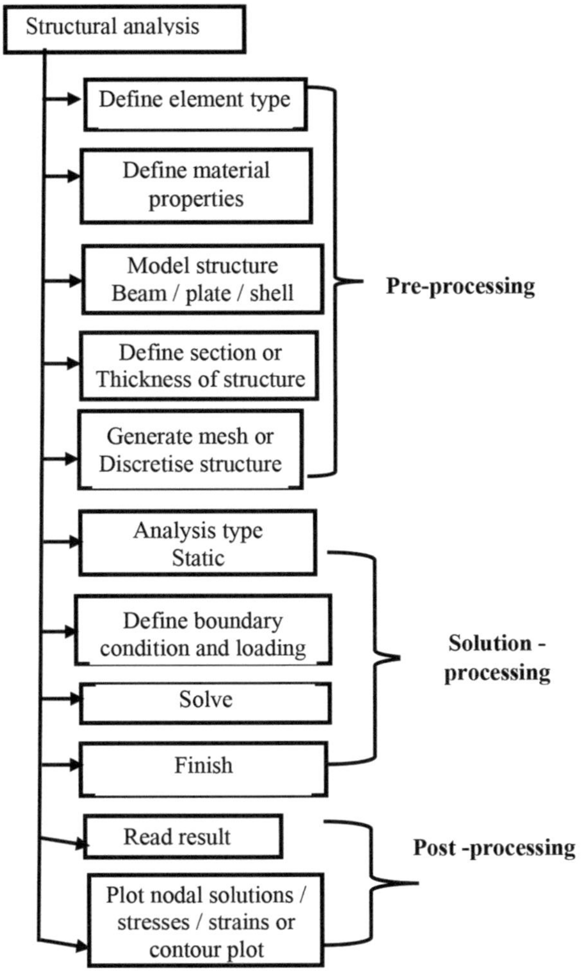

3. Results and discussion

Schematic representation of high-rise buildings considered as a vertical cantilever plate under uniformly distributed load and triangular load (wind load) is shown in **Figure 4**.

3.1 Validation study

First authors performed the validation study of developed model and then compared the present ANSYS results with available published results [28]. For comparison the deformations at specific locations ($y = 0, y = 0.25b, y = 0.5b, y = 0.75b$ and $y = b$) for a rectangular cantilever (CFFF) thin plate ($a = 12$ m, $b = 60$ m and $h = 1$ m) under uniformly distributed load ($q = 1400$ N/m^2), wind (triangular load $q \times y/b$; here $q = 1400$ N/m^2) load and combined uniformly distributed and wind load ($q/2 + (q/2) \times y/b$) of isotropic ($v = 0.3, E = 210$ GPa, $\rho = 7800$ kg/m^3) and RCC ($v = 0.2, E = 30$ GPa, $\rho = 2500$ kg/m^3) are obtained and presented in **Tables 1** and **2**, respectively.

The agreement between the present results and those from the literature is satisfactory. The method developed in this article is suitable for the problems of rectangular cantilever thin plates under uniformly distributed load, a wind load and combined uniformly distributed load and wind loads.

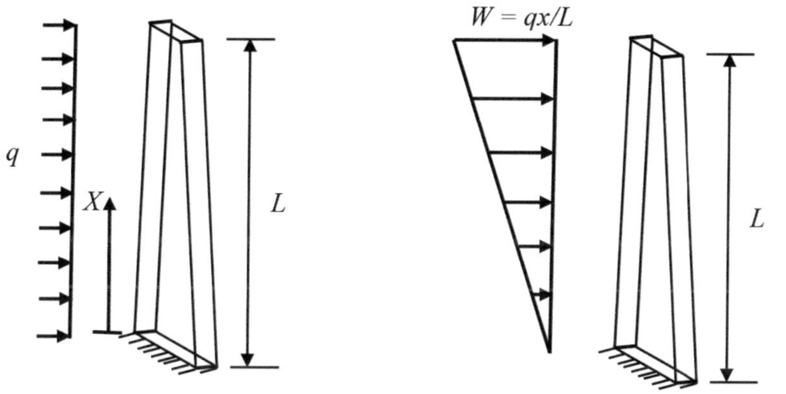

(a) Uniformly distributed load (b) Triangular load

Figure 4.
Analysis of model structure under uniformly distributed load.

δ	$y = 0$		$y = 0.25b$		$Y = 0.5b$		$y = 0.75b$		$x = b$	
	Ref [28]	**Present**	**Ref [28]**	**Present**	**Ref [28]**	**Present**	**Ref [28]**	**Present**	**Ref [28]**	**Present**
$y = 0$	0.000	0.000	0.0117	0.01394	0.0433	0.04181	0.0841	0.08361	0.127	0.1254
$y = b$	0.000	0.000	0.0117	0.01394	0.0433	0.04181	0.0841	0.08361	0.127	0.1254

Table 1.
Deformations δ (qb^4/D) [28] of the free edges x = 0 and x = a for a rectangular plate (CFFF) under uniformly distributed loading for mild steel material with $v = 0.3$.

δ	$y = 0$		$y = 0.25b$		$y = 0.5b$		$y = 0.75b$		$y = b$	
	Ref [28]	Present	Ref [28]	Present	Ref [28]	Present	Ref [28]	Present	Ref [28]	Present
y = 0	0.000	0.000	0.0117	0.01368	0.0433	0.04103	0.0841	0.0821	0.127	0.1231
y = b	0.000	0.000	0.0117	0.01368	0.0433	0.04103	0.0841	0.0821	0.127	0.1231

Table 2.
Deformations δ (Wb^4/D) [28] of the free edges x = 0 and x = a for a rectangular plate (CFFF) under uniformly distributed loading for RCC material with ν = 0.2.

3.2 Distribution of displacement, stresses and strains

Next, schematic distribution of deformation (with deformed and un-deformed shape and contour plot), membrane stresses (σ_{xx}, σ_{xy}) and second principle strain (ε_2) is shown in **Figure 5**. It is noticed that by changing the material /material properties (in terms of Young's modulus, Poisson's ratio and density) qualitatively distribution of displacement, stresses and strains are similar for cantilever towers made by isotropic material/reinforced concrete composites. In-plane minimum and maximum stresses makes a strip/band as shown in **Figure 5(d)** having magnitude 2.13 MPa and 1.53 MPa for isotropic and composite cantilever plates, respectively. Moreover, distribution of second principle strains (ε_2) is shown in **Figure 5(e),** maximum principle strains predicts at 10 m from the base represented by red color; whereas distribution of second principle strains is similar for both the cantilever plates (ν = 0.3 and ν = 0.2) under uniformly distributed load (q_0 = 1400 N/m^2).

Thereafter, the change in total displacement, in-plane strains and principle strains is shown in **Figure 6(a-e)** with schematic scales under wind load/triangular load ($q_0 \times y/b$, q_0 = 1400 N/m^2). It is observed that the distribution of longitudinal strain (ε_{xx}) and second principle strain (ε_2) is qualitatively similar as shown in **Figure 6(b** and **d)**; whereas in-plane shear strain (ε_{xy}) makes a banded strip as given in **Figure 6(c)**.

For the vibration analysis, a high-rise building was modeled as a cantilever (CFFF) plate with 12 m × 60 m with unit thickness (h = 1 m). The plate is analyzed in linear bending for displacement and vibrations using commercial software ANSYS considering two different materials- isotropic material (mild-steel young's modulus 210 GPa, poison's ratio 0.3) and RCC (young's modulus 30 GPa, poison's ratio 0.2) under three different loading conditions i.e. uniformly distributed loading, wind load and wind load with uniformly distributed loading. The results are presented in **Figures 7** and **8** for isotropic and RCC cantilever panels under various loading conditions.

Next, dynamic analysis has been performed for high-rise buildings considered as a cantilever (CFFF) plates made of isotropic (E = 210 GPA, ν = 0.3 and ρ = 7800 kg/m^3) material and RCC (E = 30 GPA, ν = 0.2 and ρ = 2500 kg/m^3) material and presented the vibration frequencies and mode shapes in **Tables 3** and **4** and **Figure 9**, respectively. The non-dimensional fundamental vibration frequencies ($\varpi_i = \omega_i b^2 \sqrt{\rho h/D}$) are given in **Table 3**, compared the present numerical results with commercial software ANSYS. Mesh convergence study is also performed to get the converge results considering mesh size 1 × 5 and 2 × 10. Non-dimensional fundamental frequencies should not change with change in material properties as given in **Table 3**. Thereafter, natural frequencies of isotropic (0.3) and composite (0.2) cantilever plates is presented in **Table 4**. Moreover, vibration mode shapes are shown in **Figure 9** representing in-plane and out-off plane bending, bending 1B, 2B, 3B; and torsion 1 T,

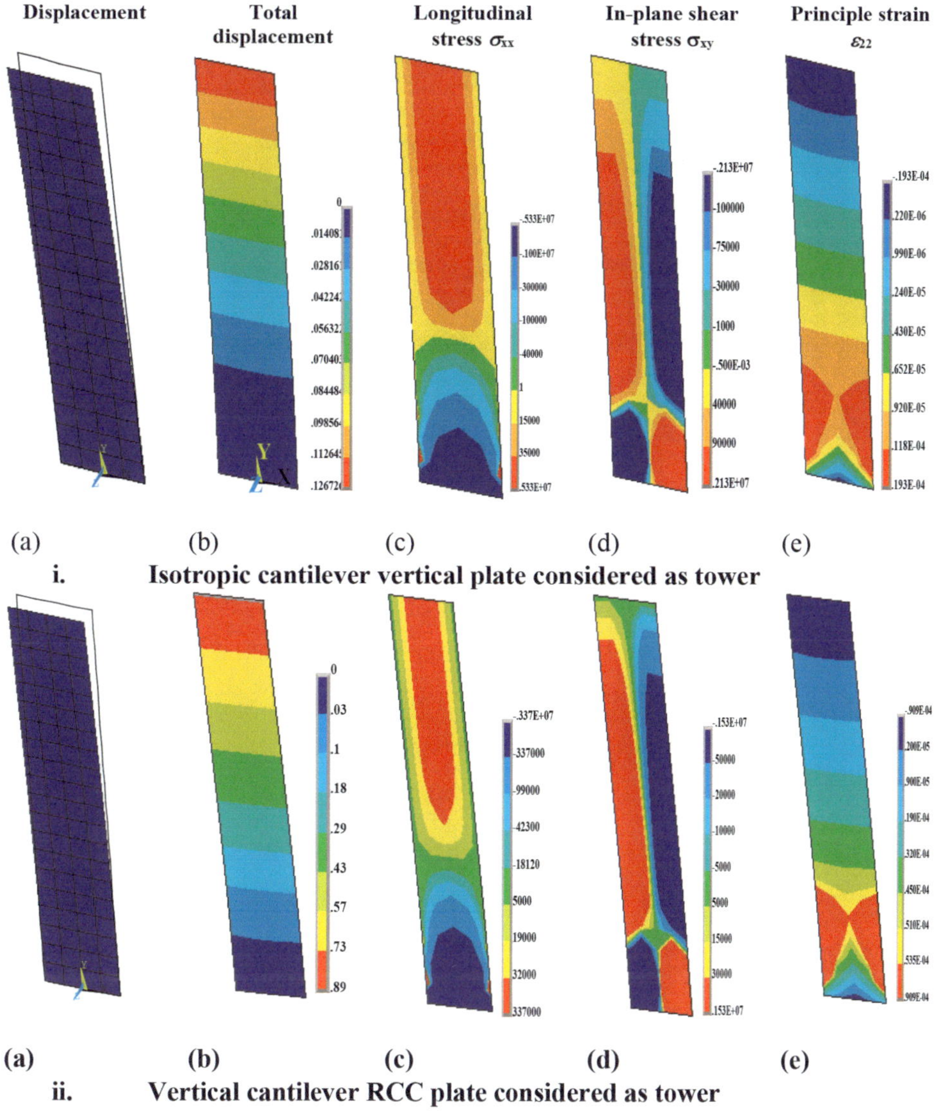

Figure 5.
Schematic representation of (a) displacement with deformed and un-deformed shape (b) Contour Plot of total displacement (c) Distribution of longitudinal stress σ_{xx} (d) In-plane shear stress σ_{xy} and (e) Second principle strain ε_2 for high rise structure considered as vertical cantilever plate ($a = 12\ m$, $b = 60\ m$, $h = 1\ m$; $E = 210\ GPa$, $\nu = 0.3$ and $\rho = 7800\ kg/m^3$) under uniformly distributed load ($q_o = 1400\ N/m^2$).

2 T, 3 T modes for mild steel and RCC cantilever plates. It is noticed that vibration mode shapes are same for isotropic and RCC structures.

From **Figures 9** and **10** it is noticed that mode shapes of Isotropic and RCC flat panel structures is same.

3.3 Large amplitude flexural vibration analysis

Next, Nonlinear vibration responses of cantilever isotropic ($\nu = 0.3$) and RCC ($\nu = 0.2$) structures is presented in **Figures 11** and **12**. It is observed that bending

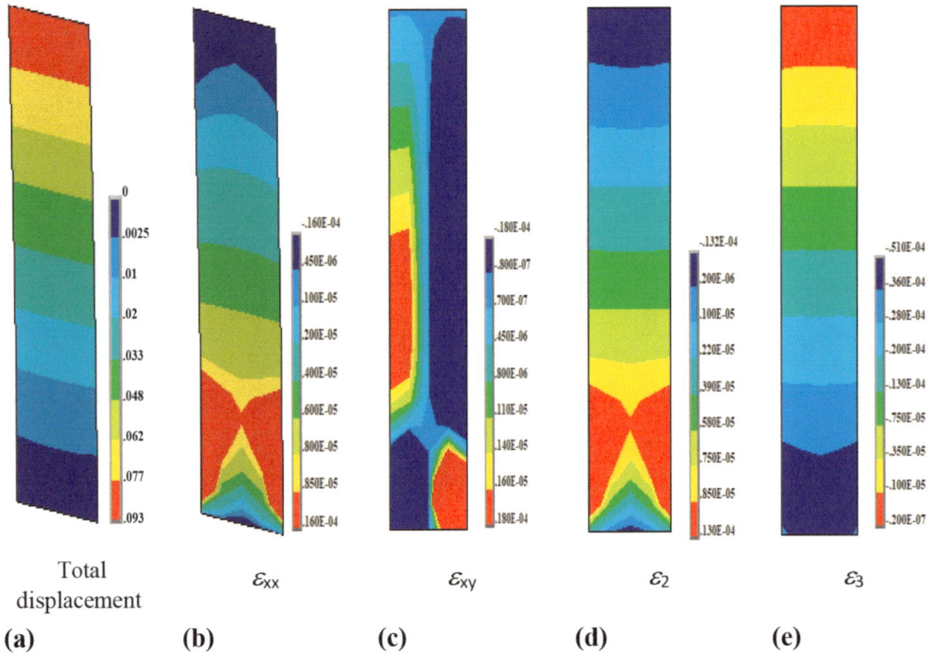

Figure 6.
Distribution of displacement and in-plane strains (ε_{xx}, ε_{xy}) and principle strains (ε_{22}, ε_{33}) under wind load ($q_o \times x/b$; q_o = 1400 N/m^2) on isotropic (ν = 0.3) cantilever plate.

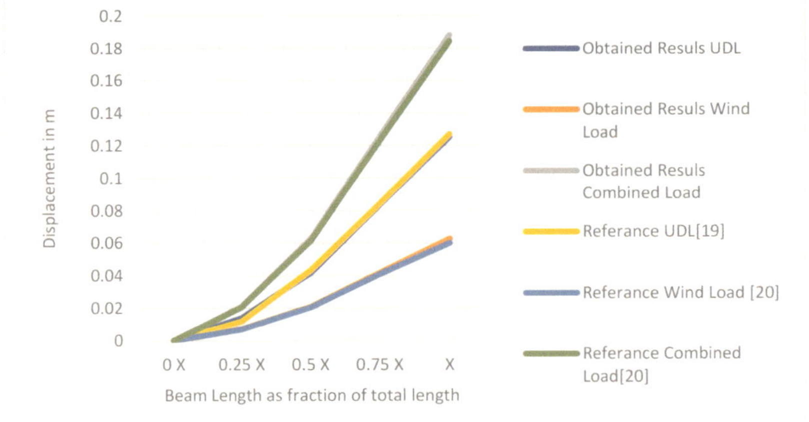

Figure 7.
Non-dimensional deformation of isotropic (ν = 0.3) thin cantilever plate.

modes (1st B, 2nd B, 1st Out-off plane B, 2nd Torsion, 3rd Bending, 3rd Torsion, 4th Bending, 4th Torsion) gives hardening response whereas 1st Torsion mode gives softening effect. Qualitatively large-amplitude flexural vibration response of isotropic and RCC structures is similar.

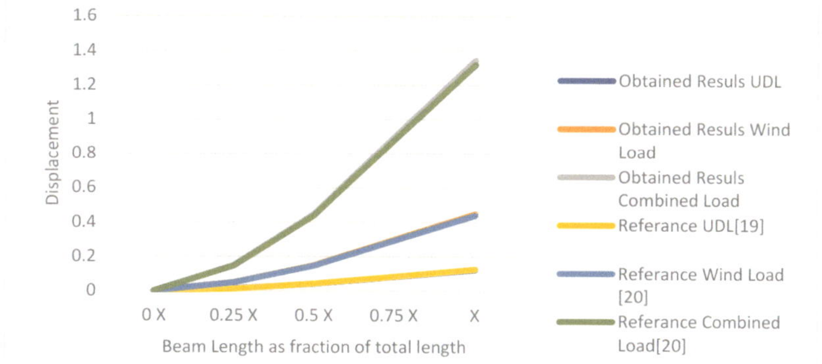

Figure 8.
Non-dimensional deformation of RCC (ν = 0.2) thin cantilever Plate.

Modes	Isotropic			RCC		
	Present		ANSYS	Present		ANSYS
	1 × 5	2 × 10		1 × 5	2 × 10	
1	3.3954	3.3930	3.3906	3.3954	3.393	3.3691
2	21.2275	21.2151	21.2000	21.2275	21.2151	21.0788
3	33.3339	33.1739	32.8655	33.3339	33.1739	34.0947
4	39.1755	39.1446	39.1403	39.1755	39.1446	39.1478
5	59.4395	59.392	59.3469	59.4395	59.3920	58.9589
6	101.9307	101.3963	100.4799	101.9307	101.3963	103.9966
7	116.6200	116.3912	116.3008	116.6200	116.3912	115.3795
8	176.0709	175.0056	173.4994	176.0709	175.0056	178.8652

Table 3.
Non-dimensional fundamental frequency ($\varpi_i = \omega_i b^2 \sqrt{\rho h/D}$) of an isotropic ($\nu$ = 0.3) and RCC ((ν = 0.2) cantilever plate (a = 12 m, b = 60 m and h = 1 m) considered as a ten story building.

Mode	1	2	3	4	5	6	7	8
Mild Steel	0.23673	1.4806	2.2944	2.7205	4.1456	7.0159	8.1275	12.119
RCC	0.16292	1.0197	1.6664	1.8929	2.8524	5.0816	5.5827	8.7343

** all values are in Hz.*

Table 4.
Natural vibration frequencies of high rise building (considered as cantilever plate).

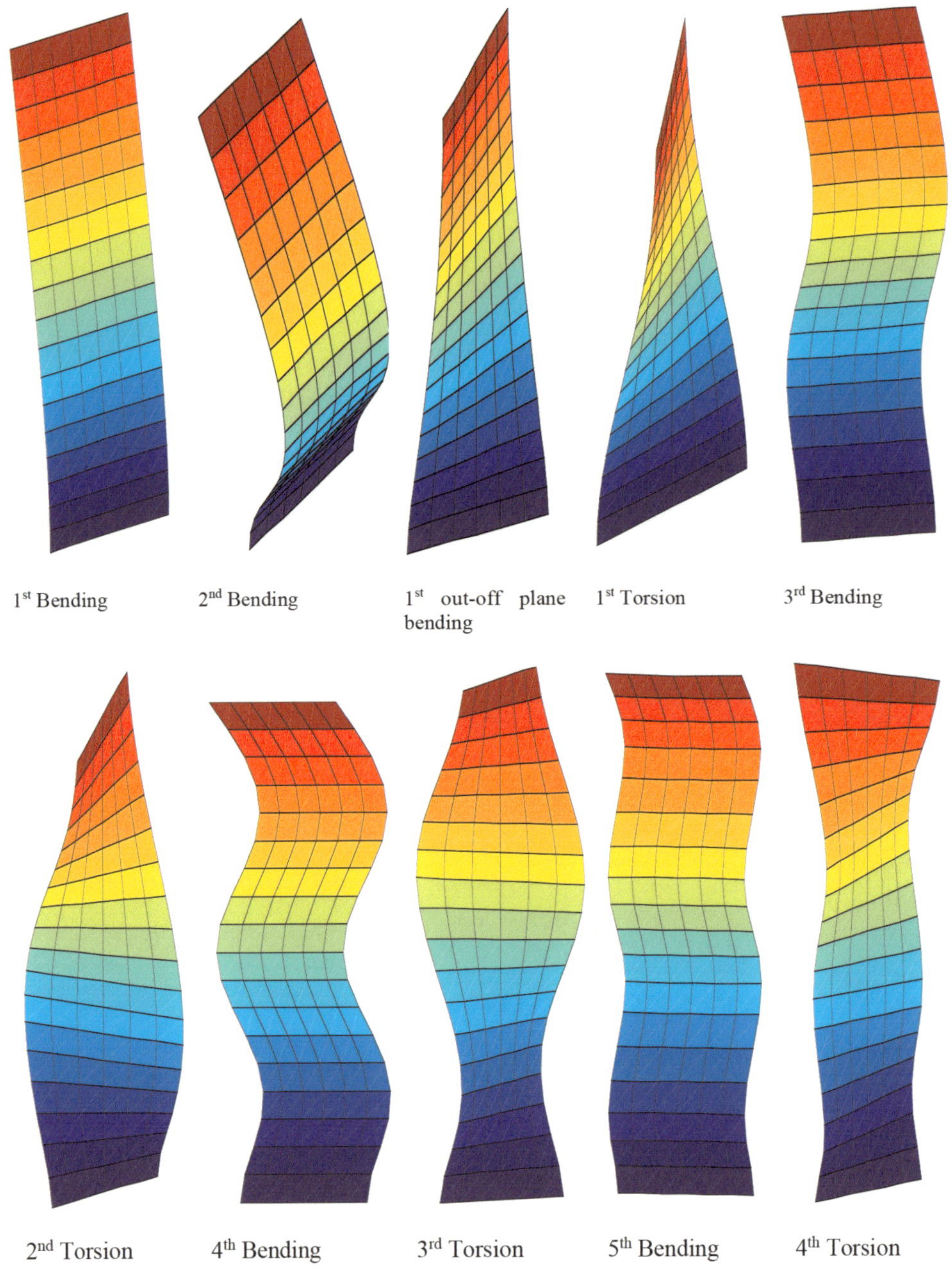

Figure 9.
Mode shapes of Isotropic and RCC cantilever plates.

3.4 Fast Fourier Transform (FFT) analysis

Next, the fast Fourier transform (FFT) and phase portraits are shown in **Figure 13** considering first three bending modes (1*B*, 2*B* and 3*B*). It is observed that non-dimensional transverse deflection (w/h) at tip of plate is maximum for first bending mode and amplitude (w/h) is reduces for higher order bending modes. Therefore, to design high-rise building first bending mode amplitude (w/h) should be minimum.

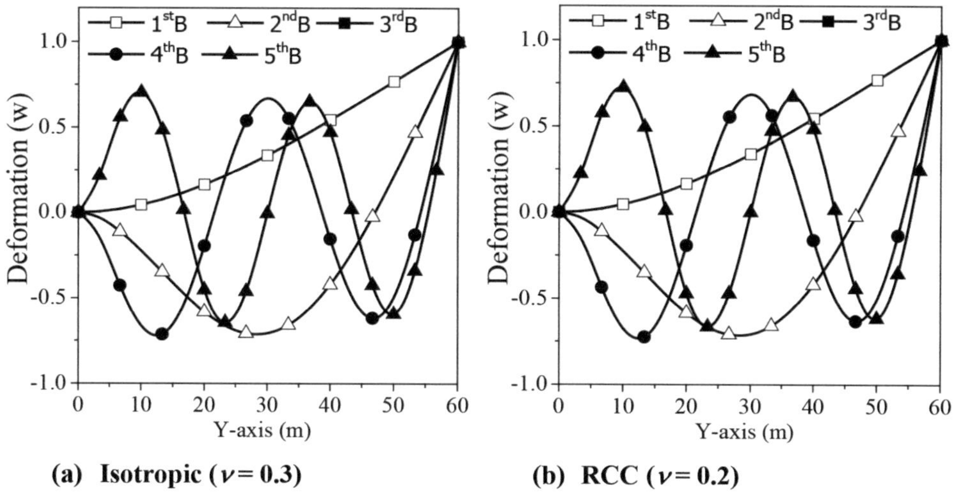

Figure 10.
Schematic representation of deformation verses y-axis for Isotropic (ν = 0.3) and RCC (ν = 0.2) cantilever plates.

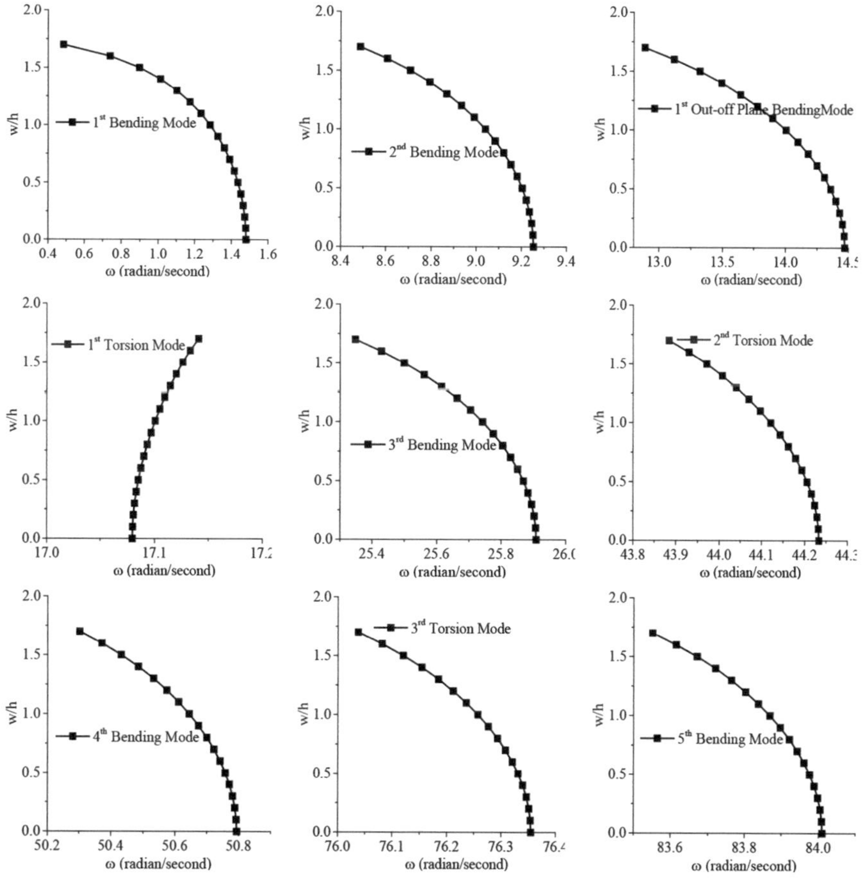

Figure 11.
The frequency (ω_i, i = 1,9) verses non-dimensional deformation (w/h) of Isotropic cantilever plate.

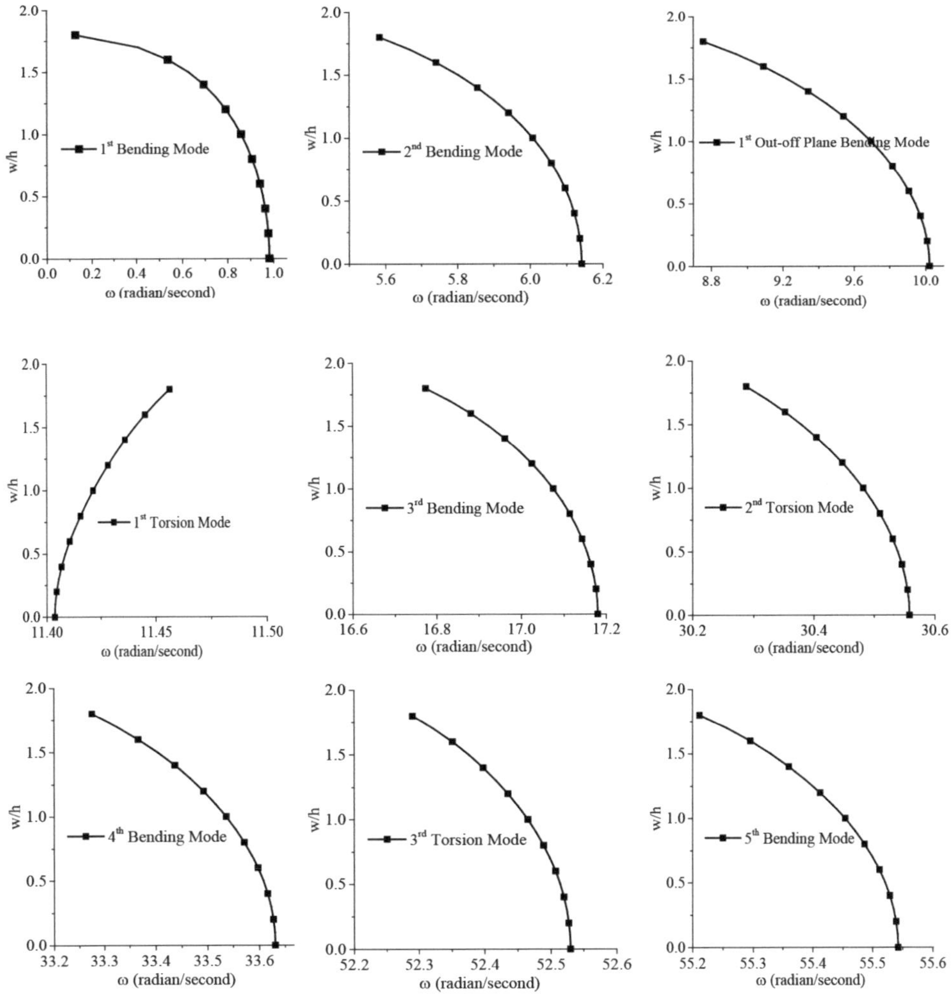

Figure 12.
The frequency (ω_i, i = 1, 9) verses non-dimensional deformation (w/h) of RCC structure.

4. Summary

In this chapter, first Authors elaborated the finite element method based mathematical formulation considering first-order shear deformation theory. Also explained the stiffness matric and compliance matrix and relationship between three dimensional stresses and strains. Authors also explained the state of in-plane stresses and in-plane stains with mathematical formulation of isotropic and anisotropic materials. Here, authors also elaborated the transformation of stiffness matrixes and explained how scaler components are independent on coordinate system and vector as well as tensors are independent, but components of vectors and tensors are dependent on coordinate system.

The static and dynamic analysis of high rise building subjected to uniform pressure (q) and wind load (considered as triangular load as $W = q \times y/L$) and combination of both ($q/2 + (q/2) \times y/L$) has been performed eight node shell 281 element through commercial finite element-based software ANSYS.

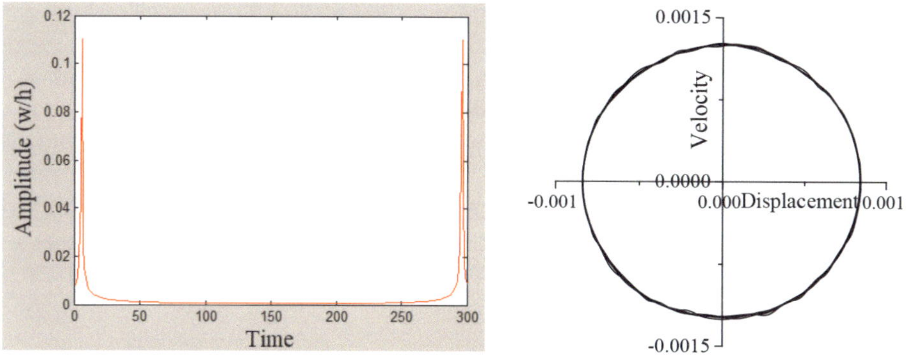

(a) 1st Bending Mode (1*B*)

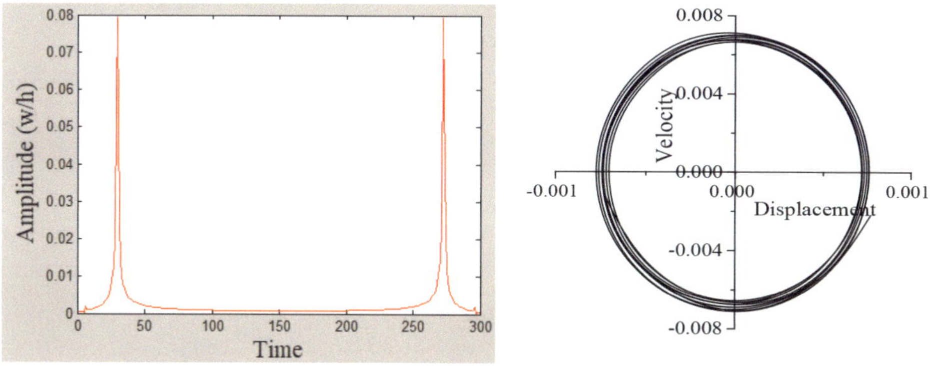

(b) 2nd Bending Mode (2*B*)

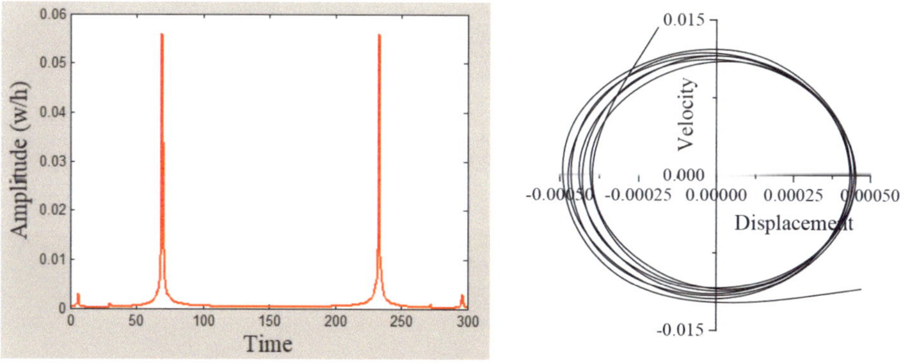

(c) 3rd Bending Mode (3*B*)

Figure 13.
Fast Fourier Transform and phase portrait for first thre`e bending modes of cantilever plate.

The large amplitude flexural vibration responses of isotropic and RCC flat panels is also investigated considering first 9 modes. It is found that 1st Torsion mode gives the softening responses whereas remaining bending, out-off plane bending and higher order torsion mode fives hardening effect.

Fast Fourier transform shows the amplitude verses time response and phase portrait gives displacement verses velocity response, represents effect mode shapes to design safe structures.

Acknowledgements

I like thank to editors of Intech Open to provide opportunity to write a book chapter to Vibration Monitoring and Analysis - Recent Advances. I also like to thank to Department of Mechanical Engineering, MBM University to provide computational facility in computer lab to work finite element method based commercial software ANSYS. I am so much thankful to editors for reviewing this chapter and help me to improve it.

Conflict of interest

The author declare no conflict of interest.

Author details

Emarti Kumari
MBM University, Jodhpur, India

*Address all correspondence to: emrati.bhaskar@gmail.com

References

[1] Pavithra SE, Vijaya S, Kavya HK. Dynamic analysis of hybrid tall tube structural system. NOVYI MIR Research Journal. 2020;**5**(12):47-54

[2] Simiu E, Yeo D. Wind Effects on Structures: Modern Structural Design for Wind. 4th ed. New Jersey, United States: Wiley-Blackwell; 2019

[3] Ankireddi S, Yang HTY. Simple ATMD control methodology for tall buildings subject to wind loads. Journal of Structural Engineering. 1996;**122**(1): 83-91. DOI: 10.1061/(ASCE)0733-9445 (1996)122:1(83)

[4] BS EN 1991-1-4:2005. Eurocode 1. Actions on Structures. General Actions. Wind Actions. London, UK: British Standards Institution (BSI); 2005

[5] Hibbeler RC, Kiang T. Structural Analysis. Upper Saddle River, NJ, United States: Pearson Prentice Hall; 2015

[6] Wen YK. Wind direction and structural reliability. Journal of Structural Engineering. 1983;**109**(4): 1028-1041. DOI: 10.1061/(ASCE) 0733-9445(1983)109:4(1028)

[7] Mehta KC, Minor JE, Reinhold TA. Wind speed-damage correlation in hurricane frederic. Journal of Structural Engineering. 1983;**109**(1):37-49. DOI: 10.1061/(ASCE)0733-9445(1983) 109:1(37)

[8] Davari SM, Malekinejad M, Rahgozar R. Static analysis of tall buildings based on Timoshenko beam theory. International Journal of Advanced Structural Engineering. 2019;**11**(4):455-461. DOI: 10.1007/s40091-019-00245-7

[9] Kwan AK. Simple method for approximate analysis of framed tube structures. Journal of Structural Engineering. 1994;**120**(4):1221-1239. DOI: 10.1061/(ASCE)0733-9445(1994) 120:4(1221)

[10] Alavi A, Rahgozar P, Rahgozar R. Minimum-weight design of high-rise structures subjected to flexural vibration at a desired natural frequency. The Structural Design of Tall and Special Buildings. 2018;**27**(15):e1515. DOI: 10.1002/tal.1515

[11] Alavi A, Rahgozar R. A simple mathematical method for optimal preliminary design of tall buildings with peak lateral deflection constraint. International Journal of Civil Engineering. 2019;**17**(7):999-1006. DOI: 10.1007/s40999-018-0349-1

[12] Dym CL, Williams HE. Estimating fundamental frequencies of tall buildings. Journal of Structural Engineering. 2007;**133**(10):1479-1483. DOI: 10.1061/(ASCE)0733-9445(2007) 133:10(1479)

[13] Bozdogan KB. An approximate method for static and dynamic analyses of symmetric wall-frame buildings. The Structural Design of Tall and Special Buildings. 2009;**18**(3):279-290. DOI: 10.1002/tal.409

[14] Kaviani P, Rahgozar R, Saffari H. Approximate analysis of tall buildings using sandwich beam models with variable cross-section. The Structural Design of Tall and Special Buildings. 2008;**17**(2):401-418. DOI: 10.1002/ tal.360

[15] Takabatake H. A simplified analytical method for high-rise buildings. Advance in Vibration Engineering and Structural Dynamics. 2012:235-283. DOI: 10.5772/51158

[16] Mohammadnejad M, Haji Kazemi H. Dynamic response analysis of tall buildings under axial force effects. Journal of Ferdowsi Civil Engineering. 2018;**31**(2):41-56

[17] Rahgozar P. Free vibration of tall buildings using energy method and Hamilton's principle. Civil Engineering Journal. 2020;**6**(5):945-953 http://dx.doi.org/10.28991/cej-2020-03091519

[18] Piccardo G, Tubino F, Luongo A. Equivalent Timoshenko linear beam model for the static and dynamic analysis of tower buildings. Applied Mathematical Modelling. 2019;**71**:77-95. DOI: 10.1016/j.apm.2019.02.005

[19] Hallebrand E, Jakobsson W. Structural Design of High-Rise. Buildings. TVSM-5000. 2016

[20] Miranda E. Approximate seismic lateral deformation demands in multistory buildings. Journal of Structural Engineering. 1999;**125**(4):417-425. DOI: 10.1061/(ASCE)0733-9445(1999)125:4(417)

[21] Miranda E, Reyes CJ. Approximate lateral drift demands in multistory buildings with nonuniform stiffness. Journal of Structural Engineering. 2002;**128**(7):840-849. DOI: 10.1061/(ASCE)0733-9445(2002)128:7(840)

[22] Su RKL, Tang TO, Liu KC. Simplified seismic assessment of buildings using non-uniform Timoshenko beam model in low-to-moderate seismicity regions. Engineering Structures. 2016;**120**:116-132. DOI: 10.1016/j.engstruct.2016.04.006

[23] Kumari E, Lal S. Nonlinear bending analysis of trapezoidal panels under thermo-mechanical load. Forces in Mechanics. 2022;**8**:1-11. DOI: 10.1016/j.finmec.2022.100097

[24] Kumari E. Dynamic response of composite panels under thermo-mechanical loading. Journal of Mechanical Science and Technology. 2022;**36**(8):1-10. DOI: 10.1007/s12206-022-04-y

[25] Kumari E. Free vibration analysis of rotating laminated composite plate type blades with variable thickness. Materials Today: Proceedings. 2021;**43**:1762-1773. DOI: 10.1016/j.matpr.2020.10.443

[26] Kumari E, Saxena D. Buckling analysis of folded structures. Materials Today: Proceedings. 2021;**43**:1421-1430. DOI: 10.1016/j.matpr.2020.09.179

[27] Naghsh A, Azhari M. Non-linear free vibration analysis of point supported laminated composite skew plates. International Journal of Non-Linear Mechanics. 2015;**76**:64-76. DOI: 10.1016/j.ijnonlinmec.2015.05.008

[28] Tian B, Zhong Y, Li R. Analytic bending solutions of rectangular cantilever thin plates. Archives of Civil and Mechanical Engineering. 2011;**11**(4):1043-1052. DOI: 10.1016/S1644-9665(12)60094-6

Chapter 6

MEMS-Based Atomic Force Microscope: Nonlinear Dynamics Analysis and Its Control

Mauricio A. Ribeiro, Jose M. Balthazar, Ângelo M. Tusset, Átila M. Bueno and Hilson H. Daum

Abstract

In this chapter, we explore a mathematical modelling that describes the nonlinear dynamic behavior of atomic force microscopy (AFM). We propose two control techniques for suppressing the chaotic motion of the system. The proposed model considers the interatomic interactions between the analyzed sample and the cantilever. These acting forces are van der Waals type, and we add a mathematical term that is a simple approximation to the viscoelasticity that possibly occurs in biological samples. We analyzed the behavior of the initial conditions of the proposed mathematical model, which showed a degree of complexity of the basins of attraction that were detected by entropy and uncertainty parameter, both detect if the basins have a fractal behavior. Numerical results showed that the nonlinear dynamic behavior has chaotic regions with the Lyapunov exponent, bifurcation diagram, and the Poincaré map. And, we propose two control techniques to suppress the chaotic movement of the AFM cantilever. First technique is the optimal linear feedback control (OLFC), which does not consider the nonlinearities of mathematical model. On the other hand, the control state dependent Riccati equation (SDRE) considers the nonlinearities of mathematical model. Both control techniques for a desired periodic orbit proved to be efficient.

Keywords: nonlinear dynamics, atomic force microscopy, control design, basins of attraction, OLFC, SDRE

1. Introduction

It is well known that many practical electromechanical devices can be modeled by a coupled equation; they can be understood in the context of simple lumped mechanical masses and electric and magnetic circuits. Electromechanical systems fall into three groups: Conventional electromechanical systems (MACRO), microelectromechanical systems (MEMS), and nanoelectromechanical systems (NEMS) [1].

Many of the "NEMS" device technologies use "MEMS" as a bridge to the nanoworld. "MEMS" will provide a bridge to enable applications of nanotechnology, as illustrated in the cantilever sensor. As an example, the atomic force microscope

(AFM) used in this technique has a tip made of silicon, using a typical MEMS device process, and is of micrometer dimensions. Cantilever sensors recognize resonant frequency shifts with the addition of mass, indicating the presence or absence of specific compounds in the environment tested. First, "MEMS" technology is based on multidisciplinary foundations. Designing a commercially viable microsystem with required dynamic performance requires an in-depth understanding and accurate prediction of its dynamic characteristics. Furthermore, the macro-properties of materials may change as the size of the feature of mechanical elements is reduced, leading to difficulties in modeling their dynamic characteristics [2].

We remark in an AFM that a microscale cantilever with a sharp tip is used to scan the specimen surface, and the vibration of the cantilever is measured to identify the distance between the tip and the specimen surface. The AFM is composed of an elastic cantilever, and the achievable sensitivity and resolution of the AFM are largely dependent on the geometry of the cantilever. Currently, AFM is one of the most effective imaging techniques that is being used at the nanoscale and sub-nanoscale levels. This technique has been applied to multiple problems in the field of natural sciences and can record a range of surface properties of materials in both liquid media and air. Nowadays, AFM includes a wide variety of methods in which the probe interacts with the sample in different ways to characterize various material properties. AFM can characterize a wide array of mechanical properties (e.g., adhesion, stiffness, friction, and dissipation), electrical properties (e.g., capacitance, electrostatic forces, work function, and electrical current), magnetic properties, and optical spectroscopic properties. In addition to imaging, the AFM probe can be used to manipulate, write, or even pull-on substrates in lithography and molecular pull experiments.

A general overview [3] was written concerning nonlinear and chaotic behavior and their controls of an atomic force microscopy (AFM) vibrating problem, which has been dedicated to tapping mode operation, considered the presence of hydrodynamic damping, based on papers of the research group in Brazil. We also discussed an AFM mathematical modeling with phase-locked loops (PLLs), inspired by Ref. [4, 5].

The MEMS-based atomic force microscope (AFM) device consists of a microcantilever beam with a tip that interacts with the surface of a sample. The sample surface topography causes vibrations in the microcantilever beam; the microcantilever reflects a laser beam that is captured by a photodiode. The laser beam deflection is used to generate the topographic images of the sample. The common operation modes of AFMs are noncontact, contact, intermittent, and trolling modes [3, 6, 7] and underserved of others. The atomic force microscope (AFM) was first presented in Ref. [8].

The field of scanning probe microscopy (SPM) began in the early 1980s with the invention of the scanning tunneling microscope (STM) by Gerd Binnig and Heinrich Rohrer, which was awarded the Nobel Prize in Physics in 1986.

Many authors developed theories on this topic. We remarked that mathematical models in (AFM) are used to analyze the behavior of strongly nonlinear dynamics, to determine the presence of irregular displacements that are caused by the interaction forces between the microcantilever tip and the atoms of the sample surface. Another contributing factor to irregular displacement in the microcantilever is the viscoelastic phenomenon. The control design are applied to suppress the irregular (chaotic) displacement of microcantilever the AFM to keep the motion regular (periodic). AFM is an essential technique for the study of surfaces and their interactions with atomic resolution, showing the strong influence of fractional nonlinear dynamics.

It is well known that the nonlinear dynamics of atomic force microscopy (AFM) is an emerging topic of research and is a widely used tool for atomic-level surface analysis. In addition, nanodevices has become an important equipment in the industry for developing nanotechnology, being used to develop substances, compounds, artifacts, and nano chips. In fact, (AFM) is a force sensor. When the surface under investigation attracts or repels the tip, the cantilever bends to or from the surface.

In this unit chapter paper, we will mention an example of a cantilever modeled as being a single spring-mass-damper system, and a nonlinear dynamic model is developed to study the cantilever-sample interaction by using the L. J. potential including the long-range attractive forces and short-range repulsive forces. A comprehensive investigation of the nonlinear dynamics and chaos is carried out based on this model, including samples related to damping; it is possible to see that the behavior is not only viscous but viscoelastic and will be discussed, using bifurcation diagrams, phase portraits, Poincare maps, and Lyapunov exponents. An active control strategy has also been proposed by us to be effective in suppressing micro-oscillations, although these methods require significant use of control resources.

Note that active methods also require sensors capable of constantly providing accurate measurements for feedback to the controller. Here, the analysis is performed considering the effects of an optimal active linear control and time-delay control.

Through computer simulations, the efficiency and robustness to parametric errors of each control technique are verified. The results obtained were in complete agreement with earlier theories and experiments. This bending is then measured by position-sensitive photodiodes via the displacement of the laser beam reflected by the back of the cantilever.

In summary, this chapter presents a short review of AFM applications of nonlinear dynamics and control. This is organized as follows. In Section 2, we presented a state of the art to give the position of the problem in the current literature. In Section 3, we exhibit the mathematical model used. In Section 4, we discuss its nonlinear dynamic behavior. In Section 5, we develop a control design to control its chaotic behavior. Finally, we present the concluding remarks and give some acknowledgments. And list the main references used.

2. Atomic force microscopy: a state of the art

A typical (AFM) system consists of a micro-machined cantilever probe and a sharp tip mounted to a piezoelectric (PZT) actuator and a position-sensitive photodetector for receiving a laser beam reflected off the endpoint of the beam to provide cantilever deflection feedback. The fundamental principle of the operation of the AFM is the measurement of the deflections of a support at the free end on which the probe is mounted. These deflections are caused by the forces acting between the probe and the sample. The effects of a variety of forces acting between the tip sample can be analyzed during the scan, as shown in **Figure 1a** (an AFM control block diagram). The diagram shows a scanned sample design, where the tip and cantilever are fixed, and the sample is moved under the tip by the piezo actuator. In this mode, the controller attempts to maintain a constant level of deflection, which corresponds to a constant level of contact force. The quantity to be measured, the surface profile, comes as an unknown disturbance to the control loop. The deflection of the cantilever is detected by optical detection. **Figure 1b** simplify the AFM schematic.

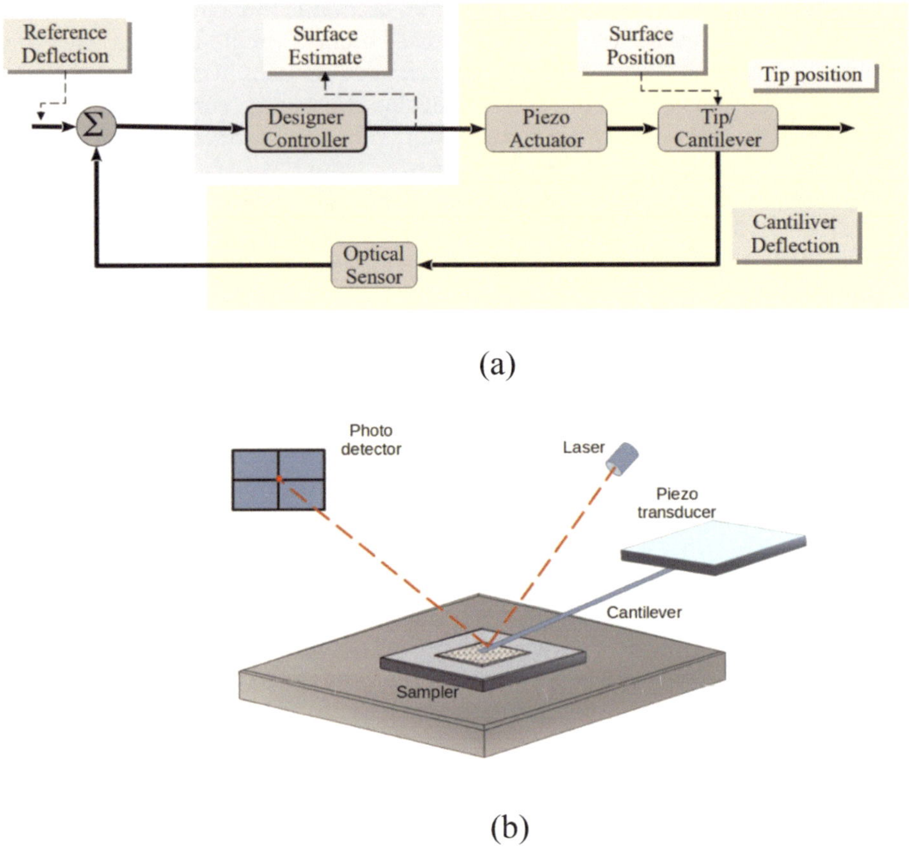

Figure 1.
(a) An AFM control block diagram and (b) AFM schematics. Adapted to [3–5, 9, 10].

Note that depending on the tip's interaction with the specimen surface, an AFM can work various imaging modes available, such as contact, noncontact, and intermittent-contact modes, tapping (where the tip oscillates and touches the surface occasionally), trolling mode (where the analysis tip is replaced by a nanoneedle that is inserted into aqueous media, the analysis splint is used in biological samples), and others. Some examples, undeserved of many others, have studied the cantilever of atomic force microscopy based on its nonlinear dynamics, listed next [3–6, 8, 9, 11–14].

The AFM microcantilever suffers from severe sensitivity degradation and noise intensification while operating in liquid; the large hydrodynamic drag between the cantilever and the surrounding liquid overwhelms the tip-sample interaction forces that are important in controlling the process. Therefore, [9] study the dynamic modeling of the manipulation process in trolling-mode AFM. The role of local and global dynamics to assess system robustness and actual safety in operating conditions is investigated, by also studying the effect of different local and global control techniques on the nonlinear behavior of a noncontact AFM. First, the nonlinear dynamical behavior of a single-mode noncontact AFM model is analyzed in terms of stability of the main periodic solutions, as well as the robustness of the attractors and the integrity of the basins [10]. The focus of the paper by Ref. [12] was on the investigation of local and global bifurcations in a continuum mechanics-based resonator model proposed for the measurement of electron spin by magnetic resonance force microscopy (MRFM).

Tapping mode AFM is one of the most potent techniques for topographic imaging of substrates. The cantilever is oscillated vertically near its resonance frequency so that the tip contacts the sample surface only briefly in each cycle of oscillation. Because of the short intermittent contact, it greatly reduces irreversible destruction of the sample surfaces, so it has been widely used for the study of soft materials, such as polymers and biological samples. When the tip is brought close to the sample surface, the vibrational characteristics of the cantilever vibration change due to the tip-sample interaction. In the imaging method, the cantilever is usually driven at the resonance frequency of the free cantilever with the driving amplitude. In Ref. [15] the authors showed how machine learning and data-driven approaches could be used to capture intermodal coupling. We employ a quasi-recurrent neural network (QRNN) for identifying mode coupling and verifying its applicability on experimental data obtained from tapping mode atomic force microscopy (AFM). The QRNN is an approach that adds convolutions to recurrence and recurrence to convolutions in the layers of the neural network to determine patterns in the system's experimental data (AFM). For details on QRNN see Ref. [16].

Accordingly, it is always required to ensure good performance of the microscope and to eliminate the possibility of chaotic motion of the microcantilever either by changing the (AFM) operating conditions to a region of the parameter space where regular motion is ensured or by designing an active controller that stabilizes the system on one of its unstable periodic orbits.

In the paper by Ref. [15], the authors investigate the mechanism of atomic force microscopy in tapping mode (AFM-TM) under the Casimir and van der Waals (VdW) force; 0–1 test was implemented to analyze the dynamics of the system, allowing the identification of the chaotic and periodic regimes of the AFM system. The dynamic results of the conventional derivative and fractional models reveal the need for the application of control techniques, such as Optimum Linear Feedback Control (OLFC), state-dependent Riccati equations (SDRE) by using feedback control, and the time-delayed feedback control. The results of the control techniques are efficient with and without the fractional-order derivative.

Ref. [16] also investigated the nonlinear dynamic model of the atomic force microscopy model (AFM) with the influence of a viscoelastic term. For the analysis of the system, we used the classic tooling of nonlinear dynamics (bifurcation diagram, 0–1 test, Poincaré maps, and the maximum Lyapunov exponent), however, the results showed the chaotic and periodic regions of the fractional system. In Ref. [17], the nonlinear dynamics and control of atomic force microscopy (AFM) in fractional order are also investigated observing the existence of chaotic behavior for some regions in the parameter space. To bring the system from a chaotic state to a periodic one, the nonlinear saturation control (NLSC) and time-delayed feedback control (TDFC) techniques for fractional order systems are applied with and without accounting for fractional order. In Ref. [18] for (AFM) fractional-order case, the results showed the influence of derivative order on the dynamics of the AFM system. Due to the fractional order, some phenomena come up, which were confirmed through detailed numerical investigations by 0–1 test. The time-delayed feedback control technique was efficient in controlling the chaotic motion of the AFM in fractional order. Furthermore, the robustness of the proposed time-delayed feedback control was tested by a sensitivity analysis of parametric uncertainties. Recently [19] considering (AFM) that the system is operating in intermittent mode, the damping dynamics of the squeeze film damping can be represented by fractional calculus through numerical simulation and dynamic analysis to prove chaotic regimes. To suppress chaotic behavior, the authors used and analyzed two control strategies, the SDRE (Riccati equation dependent states) and OLFC (Linear Control for Optimum Feedback) controls.

3. Mathematical modeling

The mathematical model analyzed is based on the atomic force microscopy system considering a viscoelastic term that is assigned to the medium in which the tip performs the analysis. For this, the proposed model considers the interatomic forces between the probe and the surface of the samples. These interatomic forces are of the van der Waals type that arises from the Lennard-Jones potential. Eqs. (1) and (2) describe the Lennard Jones potential (U_{LJ}) and the van der Waals force (F_{WD}), respectively [20, 21].

$$U_{LJ} = \frac{A_1 R}{1260 z^7} - \frac{A_2 R}{1260 z} \tag{1}$$

$$F_{WD} = \frac{-\partial U_{LJ}}{\partial z} = \frac{A_1 R}{180 z^8} - \frac{A_2 R}{6 z^2} \tag{2}$$

Where R is the radius of the tip, z is the distance of the tip, $A_1 = \pi^2 \rho_1 \rho_2 c_1$, and $A_2 = \pi^2 \rho_1 \rho_2 c_2$ are called Hamarke constants (where ρ_1 and ρ_2 are the number densities of the two interacting kinds of particles, and c_1 and c_2 are the London coefficient). Therefore, the mathematical model analyzed is the one considered by Ref. [20] in which the deflection is determined by the following:

$$w(x,t) = u(x,t) + w(x) + y(t) \tag{3}$$

where $w(x,t)$ is microcantilever beam deflection, $u(x,t)$ is a relative deflection of displacement of the actuator, described as $y(t) = Y\sin(\Omega t)$. We consider the term viscoelastic to the AFM system considering z [15, 17]:

$$F_{CS} = \frac{-\mu_{eff} b^3 L}{[z - w(L) - u(L,\ t) - Ysin(\Omega t)]^3} \dot{u} \tag{4}$$

Where μ_{eff} is the coefficient of effective viscosity, and b and L are the width and length of microcantilever, respectively. Considering the vibrations on the intermittent configuration, described by:

$$\rho A \ddot{u}(x,t) + EI(u''''(x,\ t) + w''''(x)) = \left(\frac{-A_1 R}{180(Z - w(L) - u(L,\ t) - Y\sin(\Omega t))^8} + \frac{A_2 R}{6(Z - w(L) - u(L,\ t) - Y\sin(\Omega t))^2} - \frac{\mu_{eff} b^3 L \dot{u}(x,t)}{6(Z - w(L) - u(L,\ t) - Y\sin(\Omega t))^3} \right) \times \delta(x - L) + \rho A \Omega^2 Ysin(\Omega t) \tag{5}$$

Eq. (5) is nonlinear and nonautonomous, and its discretization can be achieved through a dynamic projection on the linear modes of the system. According to Ref. [19], an approximation of the solution of Eq. (5) is by using the linear modes and frequencies of the microcantilever around its electrostatic equilibrium that are different from those of a microcantilever located far from the surface. Therefore, calculations of linear modes and microcantilever frequencies over nonlinear electrostatic equilibrium are rigorously calculated using Galerkin's method. The Galerkin's method is used to analyze problems of beams subjected to moving loads with time-varying velocities.

However, we consider under near-resonant forcing, and in the absence of additional internal resonances, only one mode of the microcantilever is assumed to participate in the response.

$$U(x,t) = \phi_1(x)q_1(t) \tag{6}$$

where $\phi_1(x)$ is the first approximate eigenfunction about the chosen equilibrium. Substitution of (6) into (5), multiplication of (5) by $\phi_1(x)$, subsequent integration over the domain, and the introduction of a modal damping consistent with the Q factors listed in **Table 1** yields the single-degree-of-freedom model:

$$\begin{aligned}\ddot{\bar{\eta}} = & -d_1\dot{\bar{\eta}} - \bar{\eta} + B_1 + \frac{C_1}{(1-\bar{\eta}-\bar{y}\sin(\Omega\tau))^8} + \frac{C_2}{(1-\bar{\eta}-\bar{y}\sin(\Omega\tau))^2} \\ & - \frac{p\dot{\bar{\eta}}}{(1-\bar{\eta}-\bar{y}\sin(\Omega\tau))^3} + \bar{y}\bar{\Omega}^2 E_1 \sin(\Omega\tau)\end{aligned} \tag{7}$$

where : $\bar{\eta} = \frac{x_1(\tau)}{\eta}$, $x_1 = \phi_1(L)q_1(\tau)$, $\eta = z - wL$, $x_1 = \phi_1(L)q_1(\tau)$, $\eta = z - w(L)$, $\tau = \omega_1 t$, $\bar{\Omega} = \frac{\Omega}{\omega_1} d_1 = \frac{C_1}{\omega_1 \rho A \int_0^L \phi_1^2 dx}$, $B_1 = (1-\bar{\alpha})\Gamma_1$, $C_1 = \frac{A_1 R}{180k(\eta)^9}\Gamma_1$, $C_2 = \frac{A_2 R}{6k(\eta)^3}\Gamma_1$, $\omega_1^2 = \frac{EI\int_0^L \phi_1 \phi_1^{\square\times} dx}{\rho A \int_0^L \phi_1^2 dx}$, $\bar{\alpha} = \frac{z}{\eta}$, $\Gamma_1 = \frac{k\phi_1^2(L)}{\omega_1^2 \rho A \int_0^L \phi_1^2 dx}$, $k = \frac{3EI}{L^3}$, $\zeta = \frac{Y}{\eta}$, $E_1 = \frac{\phi_1(L)\int_0^L \phi_1 dx}{\int_0^L \phi_1^2 dx}$, and $p = \mu_{eff} b^2 l$.

Thus, considering, $\bar{\eta} \equiv x_1$ and $\dot{\bar{\eta}} \equiv x_2$ we can rewrite Eq. (7) in the following system of differential equations:

Description	Symbol	Silicio-Silicio
Length	L	449 nm
Width	b	46 μm
Thickness	h	1.7 μm
Tip radius	R	150 nm
Material density	ρ	2330 kg m^3
Static stiffness	k	0.11 Nm^{-1}
Elastic modulus	E	176 GPa
1st resonance	f_1	11.804 kHz
Q fator (air)	Q	100
Hamaker constant (rep.)	A_1	1.3596×10^{-70} Jm^6
Hamaker constant (att.)	A_2	1.865×10^{-19} J
Coefficient of effective viscosity	μ_{eff}	$\approx 4.2101 \times 10^{-8}$ Nsm^{-2}

Table 1.
Parameters used for numerical analysis.

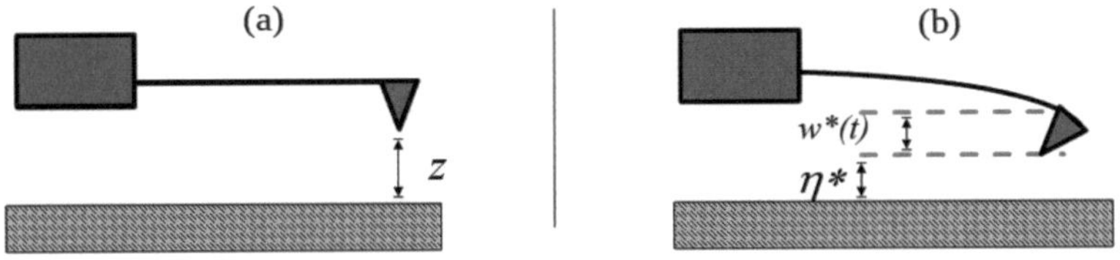

Figure 2.
Scheme of the AFM system. (a) System in initial start and (b) intermittent.

$$\begin{aligned}\frac{dx_1}{dt} &= x_2\\ \frac{dx_2}{dt} &= -d_1x_2 - x_1 + B_1 + \frac{C_1}{(1-x_1-\zeta\sin(\Omega\tau))^8} + \frac{C_2}{(1-x_1-\zeta\sin(\Omega\tau))^2}\\ &\quad \frac{-px_2}{(1-x_1-\zeta\sin(\Omega\tau))^3} + \zeta\bar{\Omega}^2 E_1 \sin(\Omega\tau)\end{aligned} \tag{8}$$

Figure 2 shows the schemes simplify of deflection behavior of nano cantilever of AFM″ for completeness.

4. Nonlinear dynamic behavior outline

The basin of attraction is the set of all starting points (initial values) that converge to the attractor. A qualitative change in the behavior (i.e., attractor) of a dynamic system is associated with a change in the control parameter. It is a qualitative leap that progresses to more complex dynamics [22–26].

Therefore, we analyzed the behavior of the basin of attraction considering the parameters $p \in [0.0.08]$ and $\eta \in [0.0.1]$. And thus determine an initial condition to perform the scan of other parameters of nonlinear dynamics, such as the maximum Lyapunov exponent (MLE) and the bifurcation diagram.

The MLE describes the divergence rate of the trajectories described by Eq. (8). The MLE has some characteristics: (i) with positive values for the Lyapunov exponent, they indicate that the trajectories of the system are divergent, this corresponds to the system having a chaotic behavior, (ii) if the MLE is negative, there is a contraction of the phase space, corresponding to a periodically stable system. For our calculations of the MLE exponent, we use the algorithm proposed by Ref. [27] with the variational calculus and the Jacobian matrix of Eq. (8).

The bifurcation diagram, in this work, is calculated considering the maximum points of the time series resulting from the integration of the system of Eq. (8) with the variation of the control parameter p. With this, we can observe the periods formed with this parametric variation. And so, to establish possible values of intervals that have multiple periods and thus diagnosing a chaotic behavior of the system, together with the MLE. And thus, obtaining the phase maps that describe the behavior of the trajectory for a given value of the parameter p.

For the dynamics calculations, we considered the parameters and properties of the cantilever described in **Table 1** for the silicon-silicon system, as seen in Ref. [28].

Table 2 are the dimensionless parameters used with the values from **Table 1** and dimensionless Eq. (8).

Parameters	Values
d_1	0.01
B_1	−0.148967
C_1	-4.59118×10^{-5}
C_2	0.149013
Ω	1.0
E_1	1.57367

Table 2.
Parameters used for numerical analysis.

4.1 Basins of attraction analysis: entropy basins and uncertainty coefficient

Suppose that we have a dynamic system with N_A attractors for a choice of parameters in a certain region Ω of the phase space. We discretize Ω through a finite number of boxes so that we cover Ω with a grid of linear size ε. Now we build an application C: $\Omega \to N$ that relates each initial condition to its attractor (which will have an associated color). Each box contains, in principle, infinite trajectories, each of which leads to an attractor labeled from 1 to N_A [22, 23].

We consider the colors in the box to be randomly distributed according to some proportions. We can assign a probability to each color j within a box i, as $p_{i,j}$ is evaluated by calculating statistics about the trajectories inside the box. Considering that the trajectories inside a box are statistically independent, we can define the Gibbs entropy of each box i is begin:

$$S_i = \sum_{j=1}^{m_i} p_{ij} \, log\left(\frac{1}{p_{ij}}\right) \tag{9}$$

Where $m_i \in [1, N_A]$ is the number of colors (attractors) in box i in $p_{i,j}$ is the probability that each color j is determined by the number of trajectories leading to that color divided by the total number of trajectories in the box. We choose nonoverlapping boxes covering Ω so that the entropy of the entire grid is calculated by adding the entropy associated with each of the N boxes given by:

$$S_i = \sum_{i=1}^{N} S_i \tag{10}$$

$$S = \sum_{i=1}^{N} \sum_{j=1}^{m_i} p_{ij} \, log\left(\frac{1}{p_{ij}}\right) \tag{11}$$

Therefore, we can consider the entropy of the basin of attraction (S_b) as follows:

$$S_b = \frac{S}{N} \tag{12}$$

An interpretation of this quantity is associated with the degree of basin uncertainty, ranging from 0 (a single attractor) to $log(N_A)$ (completely random basins with equiprobable N_A attractors). This latter higher value is rarely realized in practice, even for extremely chaotic systems. In some cases, we may only be interested in the uncertainty of boundaries between basins of attraction. We often want to know if the boundary is fractal. For this purpose, we can restrict the calculation of the basin entropy to the boxes that fall within the boundaries of the basin of attraction. We can calculate the entropy only for the N_b boxes that contain more than one attractor (color),

$$S_{bb} = \frac{S}{N_b} \tag{13}$$

where S is defined by Eq. (11). We refer to this S_{bb} number as the basin entropy quantifies the uncertainty regarding the boundaries only. The nature of this S_{bb} quantity is different from the entropy of the Sb basin, since Sb is sensitive to the size of the basins, so it can distinguish between different basins with smooth boundaries, S_{bb} provides a sufficient condition to easily assess that some boundaries are fractals [22–24].

Another way to quantify this uncertainty in the initial conditions for its final state is through the uncertainty coefficient. The uncertainty coefficient is related to the sensitivity of the final state of the trajectories in the phase space. An exponent close to 1 means that the basin has smooth contours, while an exponent close to 0 represents fully fractalized basins, also called sieve basins [26].

A phase portrait with a fractal boundary can cause uncertainty in the final state of the dynamical system for a given initial condition. To determine the uncertainty coefficient, one must probe the basin of attraction with balls of size ε at random. If there are at least two initial conditions that lead to different attractors, a ball is marked "*uncertain.*" In this way, we can denominate the fraction of "uncertain balls" (f_ε) for the total number of attempts in the basin. In analogy to the fractal dimension, there is a scaling law between, $f_\varepsilon \sim \varepsilon^\alpha$. The number that characterizes this scale is called the uncertainty exponent α [26]. For our analysis, we considered the set of differential equations that describe the interactions of the atomic force microscopy system and the parameters described in **Table 1**.

For this we consider $p \in [0,0.08]$ and $\eta \in [0,0.1]$, and for the numerical analysis of S_b, S_{bb}, and α we use an interval of initial conditions $x_1^0 \times x_2^0 = [-0.9,0.9] \times [-0.9,0.9]$ with approximately 250000 initial conditions and making a 200 x 200 grid considering ε=0.002 for each attraction basin formed during the numerical analysis. **Figure 3a** shows the behavior of S_b, **Figure 3b** shows behavior of S_{bb}, and **Figure 3c** shows behavior of α for set parameter $p \in [0,0.08]$ e $\eta \in [0,0.1]$.

We can see in **Figure 3a** and **b** that the red regions show the maximum values for S_b and S_{bb}, that is, the regions where the basins of attraction have more attractors and

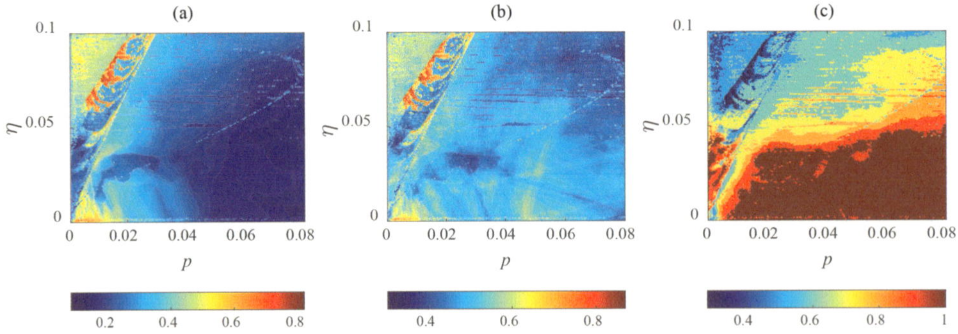

Figure 3.
(a) shows the behavior of S_b, (b) shows behavior of S_{bb}, and (c) shows behavior of α for set parameter $p \in [0,0.08]$ e $\eta \in [0,0.1]$.

	p	η
S_b	0.0012	0.0045
S_{bb}	0.0012	0.0045
α	0.0764	0.0121

Table 3.
Summarizes the parameters p and η that provide the maximum. values for S_b and S_{bb} and the minimum value of α that produce the basins of attraction.

their edges are fractalized. However, for the uncertainty coefficient in **Figure 3c** the uncertainty coefficient is close to 1 showing the smooth basins. According to Ref. [20] the difference between S_b and S_{bb} and the uncertainty coefficient of the attraction basins is that when S_b and S_{bb} are minimum the uncertainty coefficient is maximum, or when S_b and S_{bb} are maximum the uncertainty coefficient is minimum. This corroborates the analysis of the behavior of the initial conditions with the parametric variation of p and η. **Table 3** shows the parameters p and η that provide the maximum values for S_b and S_{bb} and the minimum value of α that produces the basins of attraction.

Figure 4a–c show the behavior of the attraction basins considering the maximum values of S_b and S_{bb} and for the minimum value of α.

Figure 5a–c represent attractor points referring to the basins of attraction of **Figure 5a–c**.

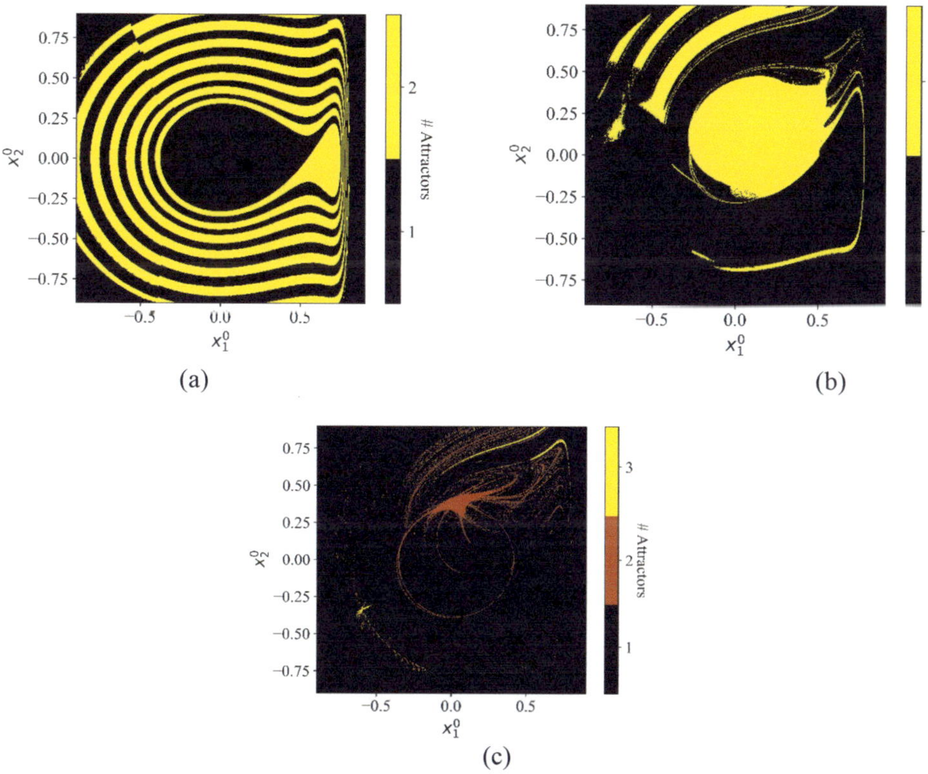

Figure 4.
Basins of attraction. (a) p = 0.0012 and η = 0.0045, (b) p = 0.0012 and η = 0.0045, and (c) p = 0.0764 and η = 0.0121.

4.2 Numerical dynamics analysis

Taking into account the analysis of the basins of attraction for the interval of $p \in [0,0.08]$ and $\eta \in [0.0.25]$, we adopted a larger interval for η, for analysis of the maximum Lyapunov exponent (MLE) and the initial condition [0.1, 0.0], because depending on the values of p and η the initial condition can participate in different attractors, as we saw in **Figures 4a** and **b** and **5a** and **b**.

Using the Jacobian matrix for the variational calculus and the sweep of $p \in [0,0.08]$ and $\eta \in [0,0.25]$, we have the behavior of the MLE. **Figure 6** shows the space of MLE parameters in which the region of white to black shows the regions in Eq. (7) has a periodic behavior. However, for the region between yellow and green, it shows the chaotic behavior with the parameter sweep $p \times \eta$.

Figure 7a shows the behavior of the bifurcation diagram for the following parameters, so we can observe the periodic windows $p \in [0, 0.0777]$ (black region), the intervals are confirmed by the maximum Lyapunov exponent. However, for the interval $p \in [0.0777, 0.08]$ there is a chaotic window (red region).

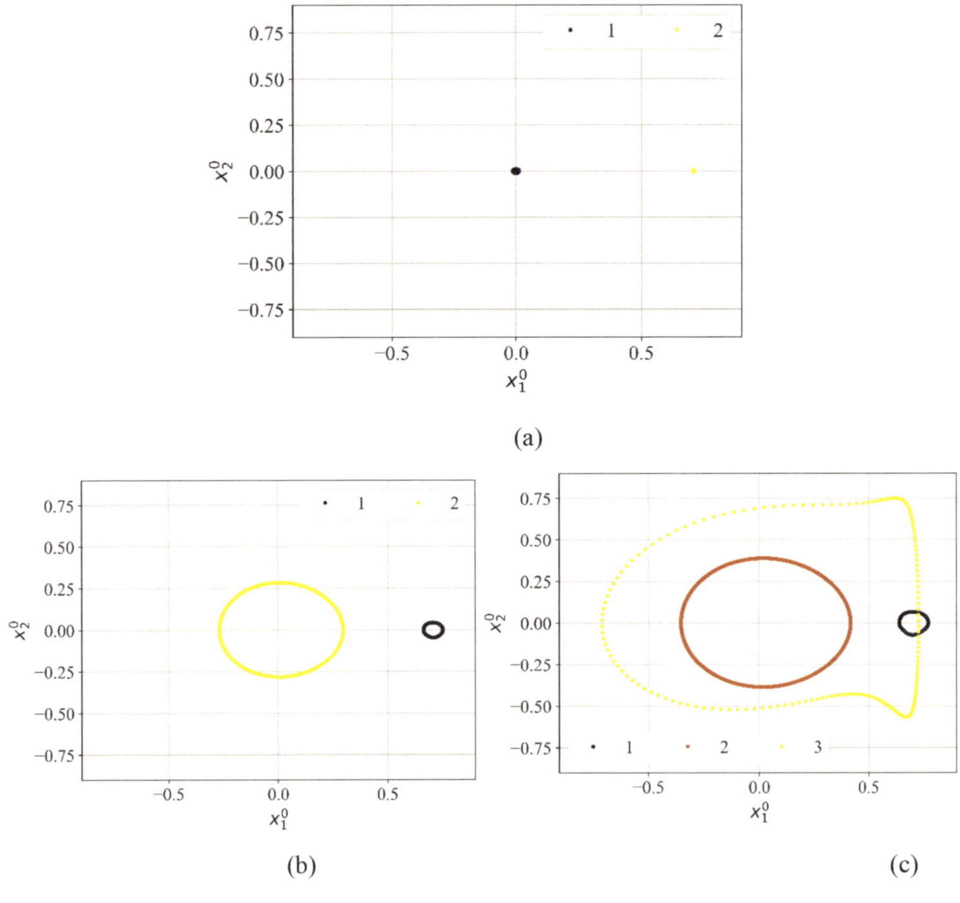

Figure 5.
Attractive orbits that form the basins of attraction of the figure. (a) p = 0.0012 and η = 0.0045, (b) p = 0.0012 and η = 0.0045, and (c) p = 0.0764 and η = 0.0121.

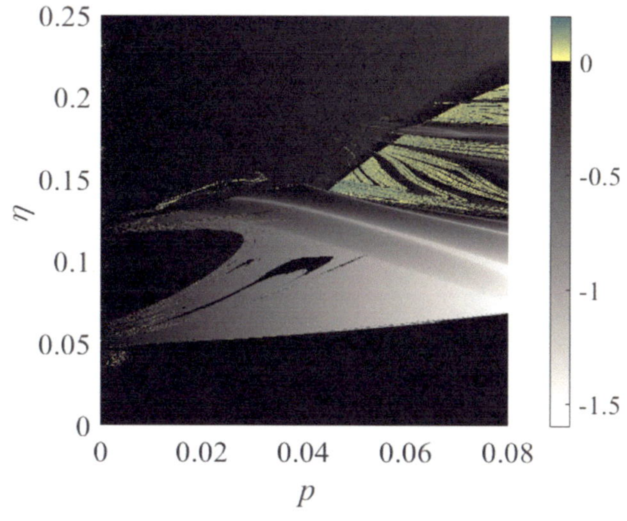

Figure 6.
Behavior of the maximum Lyapunov exponent (MLE) for the parameters $p \in [0, 0.08]$ *and* $\eta \in [0, 0.25]$.

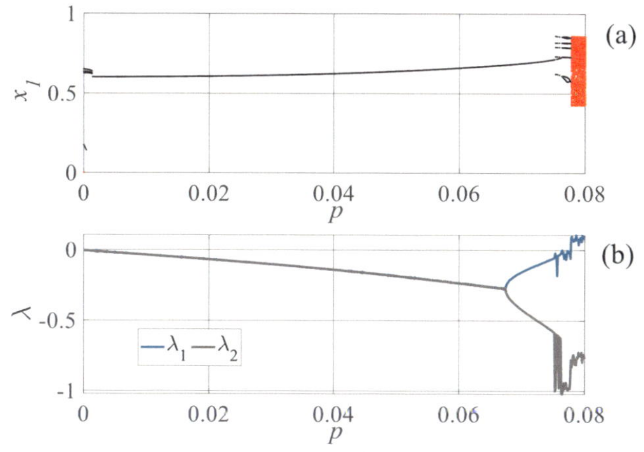

Figure 7.
(a) Bifurcation diagram for displacement x_1 *in black region shows the periodic behavior, red region shows the chaotic behavior, and (b) Lyapunov exponent.*

Therefore, considering the previous nonlinear dynamic results, we obtain the phase portrait and the Poincare map for p =0.0791 and η = 0.2043 (**Figure 8**). **Figure 9a** shows the phase portrait (Gray Line) and Poincare map (Black Dots). **Figure 9b** and **c** show the time series of displacement x_1 and velocity x_2.

5. Control design

The control techniques are Optimal linear feedback control (OLFC) and state-dependent Riccati equation (SDRE).

The OLFC has control feedback as its main application for controlling nonlinear systems, it has this name because it is characteristic of the controller to have the control

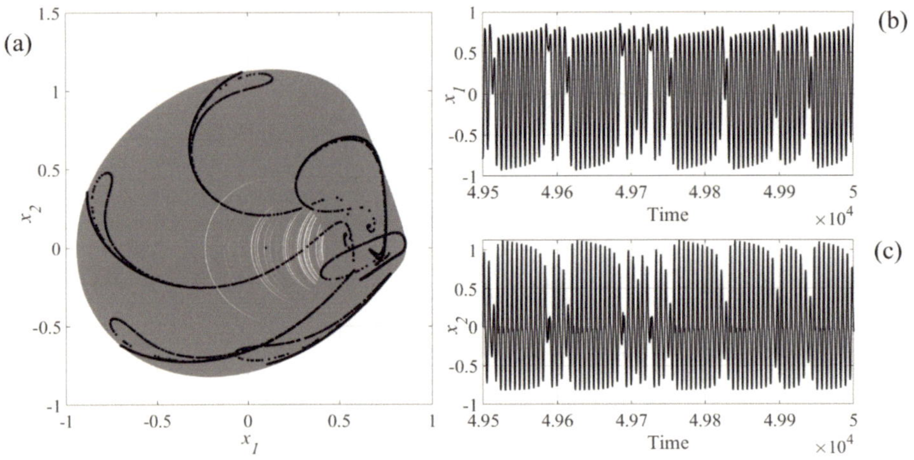

Figure 8.
(a) Phase Portrait (gray lines) and Poincare Map (black dots), (b) time series of displacement x_1 *and (b) velocity* x_2*.*

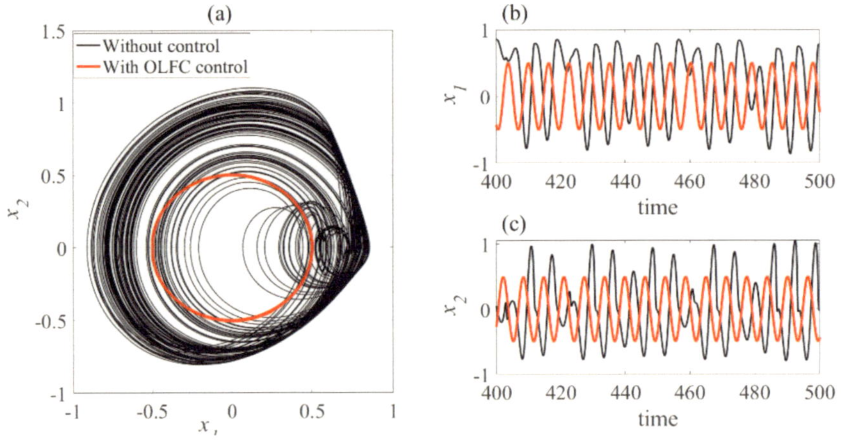

Figure 9.
(a) Portrait phase (black line) and orbit (red line), (b) time series of displacement x_1*, and (c) and time series of velocity* x_2*.*

signal as a function of the difference between real values and the values expected by the state variables. The application of this technique in nonlinear systems is due to the simplicity of implementation. In other words, the OLFC control does not consider the nonlinearities of the system of equations for the suppression of the chaos that occurred in the microcantilever of the AFM system. However, the SDRE controller considers the nonlinearities of the system, and its state matrix is not fixed. The SDRE control technique guarantees an asymptotically stable solution over the origin [7, 29–33].

In both cases, there is a need to propose an orbit for the system to be controlled. In this chapter, we opted for simplicity of calculations to consider an orbit described by Eq. (14):

$$\begin{cases} x_1 = 0.5 sin\,(\omega t) \\ x_2 = 0.5 cos\,(\omega t) \end{cases} \tag{14}$$

We can consider the system Eq. (8) in the matrix form given by:

$$\dot{x} = Ax + g(x) + U \tag{15}$$

Where **Ax** is error dependent state matrix, **g(x)** is a nonlinear matrix, and **U** is control signal [34, 35].

Both control techniques use two controllers called feedback (u_f) and feedforward (ũ), so the control signal of the nonlinear system is defined **U**=ũ+u?. While u_f has the characteristic of correcting the difference between the real values and the stipulated values, taking the system to the stipulated orbit, ũ has the purpose of keeping the system in the desired orbit [24, 25]. The SDRE controller design, like the OLFC, follows some steps to obtain the optimal solution to the dynamic control problem [10, 31]:

- Define the state space of the model and parameterize the model in the form of state-dependent coefficients.
- Measures the state of the system x(t), that is, to define x (0) = x_0, and to choose the coefficients of the matrices **Q** and **R.**
- Solve the Riccati equation for the state x(t);
- Calculate the input signal of the state feedback control equation.
- Check controllability
- Assume the output value of the system as a new initial value and update the state of the system x (t). Recalculate the Riccati equation and repeat the process until the defined stopping criterion is reached.

5.1 Optimal linear feedback control OLFC

The optimal linear feedback control is used in nonlinear systems due to its simplicity in its implementation since the control uses fixed k gains. The value of **K** is obtained by solving the Riccati equation [34–36].

The nonlinear system of Eq. (7) can be written in matrix form using the general equation of nonlinear systems 14.

Then Eq. (16) shows the system rewritten in matrix form and as a function of errors, knowing that $= (X - \tilde{X})$, then it is possible to obtain the state matrix A error-dependent and nonlinearity matrix $g(e, \tilde{X})$,

which will not be used in the control and, therefore, is indicated in the equation.

$$\dot{e} = \begin{pmatrix} 0 & 1 \\ -1 & D1 \end{pmatrix} \cdot \begin{pmatrix} e_1 \\ e_2 \end{pmatrix} + g(e, \tilde{X}) + U \tag{16}$$

The gain is described by$k = R^{-1}B^TP$ where P is the Riccati matrix, $R=\begin{pmatrix} 1 & 0 \\ 0 & 1 \end{pmatrix}$,

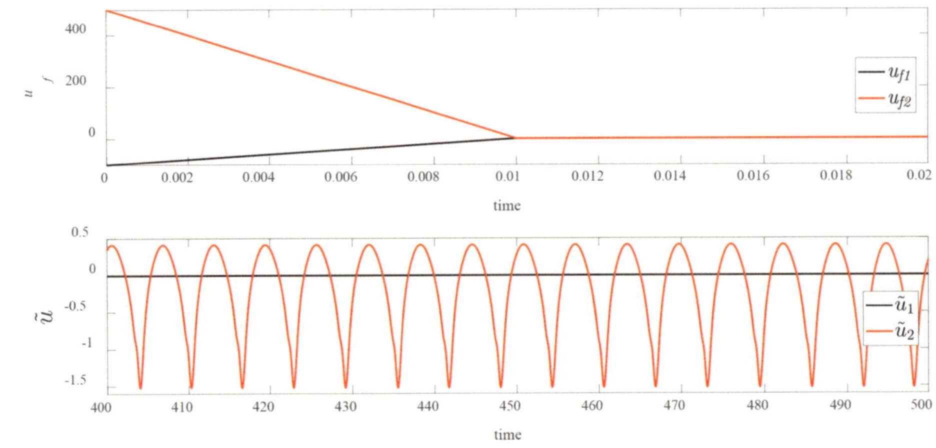

Figure 10.
u_f in the transient regime and $\tilde{u}$ is the steady-state control signal.

$B=\begin{pmatrix} 1 & 0 \\ 0 & 1 \end{pmatrix}$, and $Q=1000 * \begin{pmatrix} 1 & 0 \\ 0 & 1 \end{pmatrix}$ [23, 24, 33].

The matrices P,R,B, and Q guarantee the stability of the solution of the Riccati equation. Therefore, the gain found by the system is defined by

$$k = \begin{pmatrix} 1000 & 5 \times 10^{-6} \\ 5 \times 10^{-6} & 1000 \end{pmatrix} \tag{17}$$

Figure 9a shows the portrait phase of the uncontrolled system in black and controlled in red, and (b) and (c) the time series of the system is shown.

Figure 10 shows the control signal $u_?$ in the transient regime, that is, at the moment when the control is taking the system to the desired orbit in the plot of ũ is the steady-state control signal.

In **Figures 11** and **12,** the controller errors in transient and steady state are presented for each time series of the system.

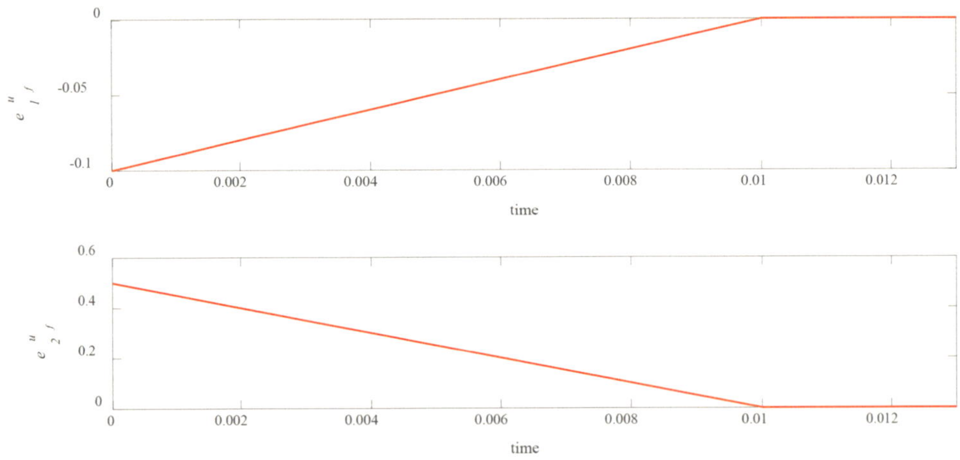

Figure 11.
$e_1^{u_f}$ represents the feedback control and $e_2^{u_f}$ represents the feedback control.

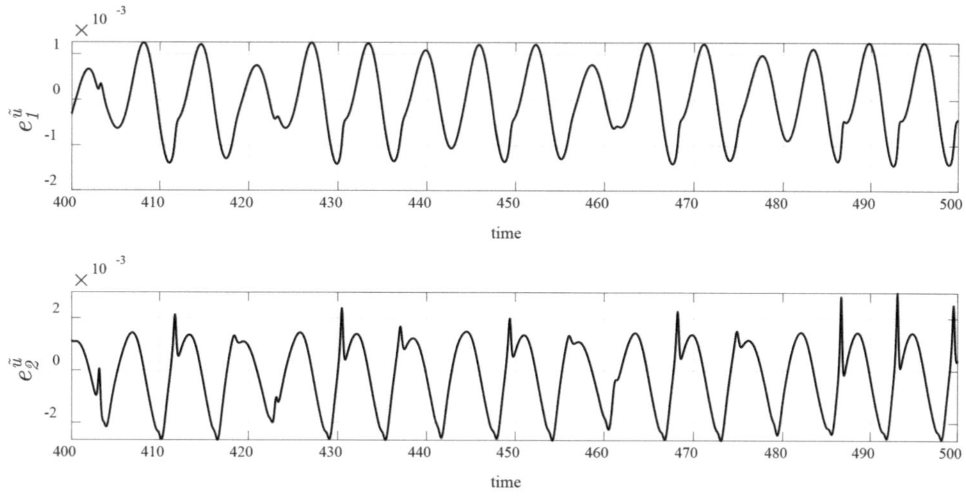

Figure 12.
$e_1^{\tilde{u}}$ represents the error of x_1 in steady state for the feedforward control. $e_2^{\tilde{u}}$ represents the trajectory error of x_2 in the control feedforward.

5.2 State-dependent Riccati equation control

The state-dependent Riccati equation (SDRE) control, unlike the OLFC, does not exclude the dependence of the error-dependent state matrix, so the controller gains change with each iteration. The SDRE methodology used to find error-dependent states used matrix like that is used for the OLFC control. It can be written from the following nonlinear matrix of errors:

$$\dot{e} = \left(\frac{2c2}{(1-x_1-\eta_1 \sin(\omega t))^3} - \frac{\overset{0}{3x_2p}}{(1-x_1-\eta_1 \sin(\omega t))^4} + \frac{8c1}{(1-x_1-\eta_1 \sin(\omega t))^9} \quad \frac{\overset{1}{p}}{(1-x_1-\eta_1 \sin(\omega t))^3} - d_1 \right) \cdot \begin{pmatrix} e_1 \\ e_2 \end{pmatrix} + g(\tilde{X}) + U \tag{18}$$

where $g(\tilde{X})$ is the matrix that does not depend on errors. The gain **k** is obtained $k = R^{-1}B^TP$ where P is the Riccati matrix. The control $\boldsymbol{u}$ found from the solution of the following equation:

$$u = -R^{-1}B^TPe \tag{19}$$

Being a symmetric matrix and obtained from the algebraic Riccati equations [7, 34, 36]:

$$A^TP + PA - PBR^{-1}B^TP + Q = 0 \tag{20}$$

The controller gain k is defined $k = R^{-1}B^TP$ where P is the Riccati matrix, $R=\begin{pmatrix} 1 & 0 \\ 0 & 1 \end{pmatrix}$, $B=\begin{pmatrix} 1 & 0 \\ 0 & 1 \end{pmatrix}$, and $Q=1000\begin{pmatrix} 1 & 0 \\ 0 & 1 \end{pmatrix}$ [23, 24, 33].

The matrices P,R,B, and Q guarantee the stability of the solution of the Riccati equation. Therefore, the gain found by the system is defined in **Figure 13,** which

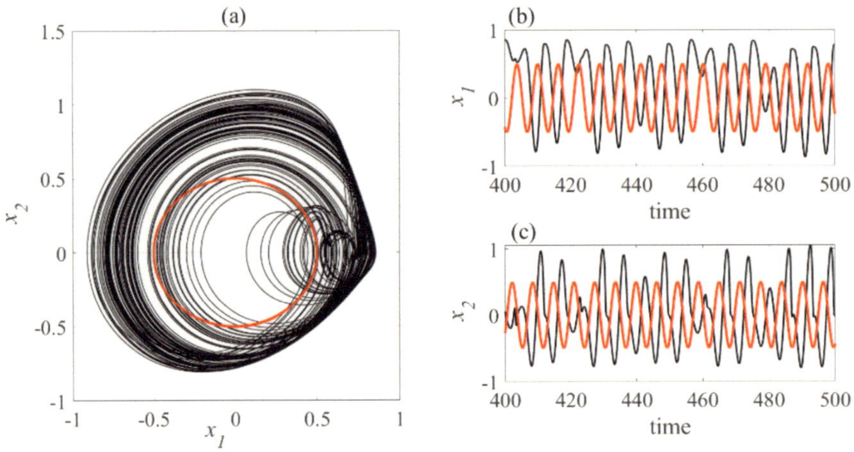

Figure 13.
(a) Portrait phase (black line) and orbit (red line), (b) time series of displacement x_1, *and (c) time series of velocity* x_2.

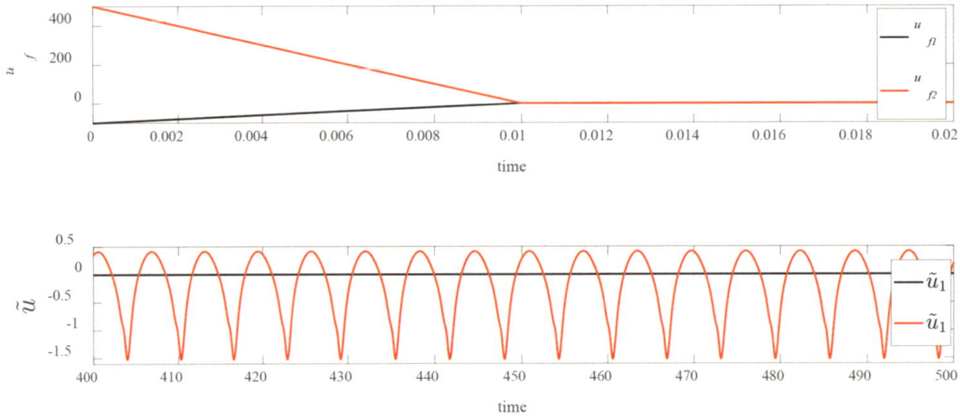

Figure 14.
SDRE signal control.

shows the result of the controlled system using the SDRE technique, in **Figure 13a** the portrait phase without control and with control is presented, and their respective time series in **Figure 13b** and **c**.

Figure 14 represents the control signal u_f in the transient regime, that is, now when the control is taking the system to the desired orbit. The $\tilde{\mathbf{u}}$ plots show the steady-state control signal.

Figures 15 and **16** show the controller errors in transient and steady state for each time series of the system.

Figure 17 shows the behavior of gain $\boldsymbol{k}$ since it is calculated at each interaction, that is, the controller gain is variable.

Figure 18 shows the comparison of the control signal $\boldsymbol{U}=\tilde{\boldsymbol{u}}+\boldsymbol{u_f}$ for the two techniques applied in this text. It is possible to notice that the OLFC control, even excluding the nonlinearities of the system, has a control signal very close to the control signal of the SDRE.

DOI: http://dx.doi.org/10.5772/intechopen.108880

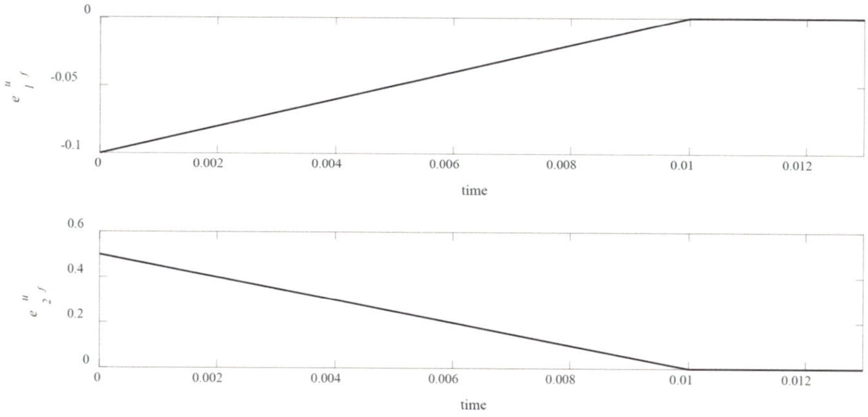

Figure 15.
$e_1^{u_f}$ represents the error of x_1 for the feedback control and $e_2^{u_f}$ represents the path of x_2 for the feedback control.

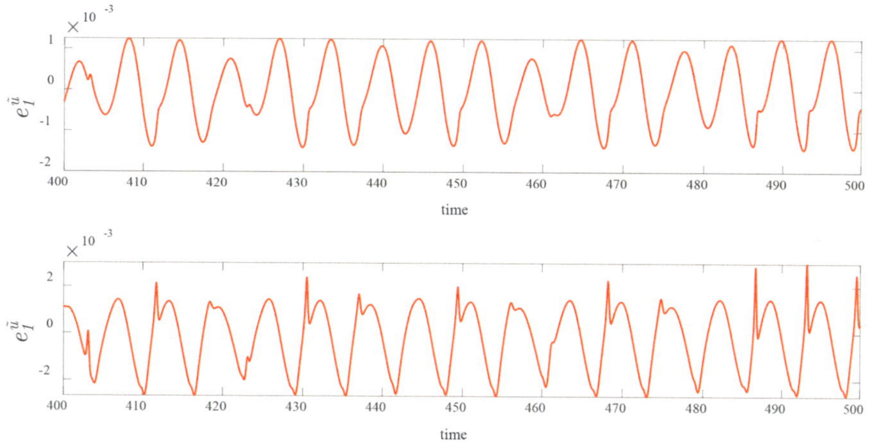

Figure 16.
$e_1^{\tilde{u}}$ represents the error of Eq. (7) in steady state for feedforward control. $e_2^{\tilde{u}}$ represents the steady-state trajectory error for the feedforward control.

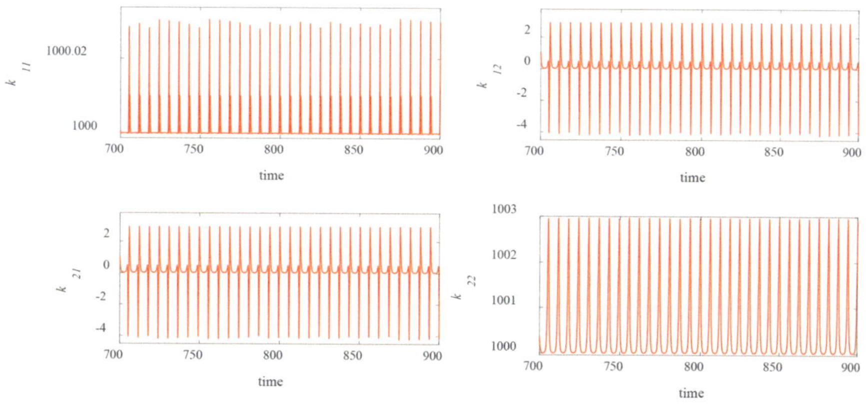

Figure 17.
*SDRE behavior of gain **k**.*

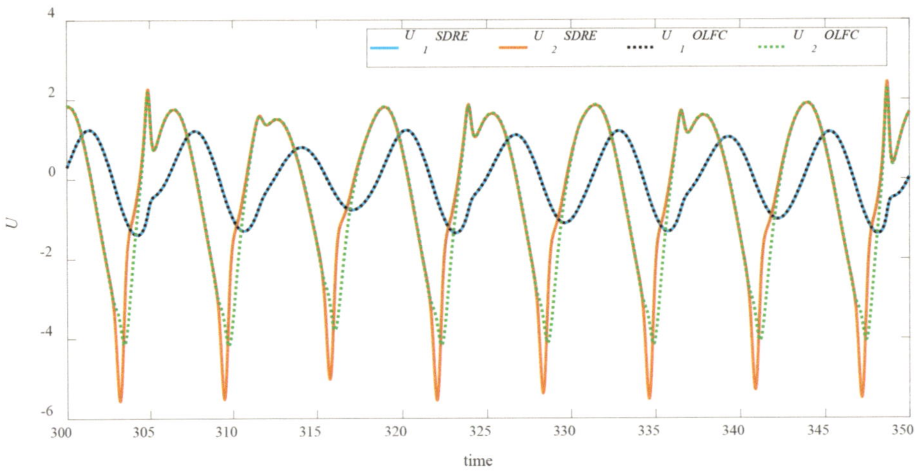

Figure 18.
Comparison of the control signal.

6. Conclusions

In this chapter, we describe the applications of atomic force microscopy. We also analyzed the nonlinear dynamic behavior of a mathematical model considering the viscoelasticity term and the microcantilever deflection. In this way, we establish the behavior of the initial conditions and which ones have greater entropy and more fractality. These analyzes corroborate to determine the set of initial conditions for our dynamic analyses, as observed in **Figure 3** and **Table 3**.

After analyzing the behavior of the initial conditions, the dynamic behavior of the dimensionless parameter was analyzed, which considers viscoelasticity and, therefore, the regions in which the system presents a chaotic behavior. This behavior was obtained using the maximum Lyapunov exponent, and for a given set of parameters, it was observed by the bifurcation diagram and the Poincaré map. In this way, the ranges for the parameters p and η were established where a possible chaotic behavior occurs, as we see in **Figures 5** and 7.

The results obtained by the analysis of the basins of attraction showed a strong influence between the parameters p and eta in the initial conditions. As we observed in the calculation of entropy and uncertainty coefficient for the grid of initial conditions $x_1^0 \times x_2^0 = [-0.9,0.9] \times [-0.9,0.9]$, have regions of high fractality and receive it from a new attractor, as shown in **Figures 4a–c** and **5a–c**. Considering the initial condition $[0.1,0.0]$ the MLE has regions of positive value, that is, there is a chaotic behavior, as shown in **Figure 6**.

These analyses corroborate to determine the p and η parameters for the application of two control techniques and suppress the chaotic behavior. This suppression allows us to have a better understanding of the microcantilever when reading biological samples that can generate chaotic movements. These chaotic movements can be detected as noises in the structure of the AFM device.

In general, it is possible to notice that the two control methods presented low errors as shown in **Figure 16**. For this system there was no difference in convergence to the intended orbit; however, it is possible to notice that the OLFC control has a simple implementation methodology in relation to the SDRE, as it excludes nonlinearities, facilitating the application of the control method and enabling a possible practical implementation using embedded systems. **Figure 17** showed this comparison between the control techniques.

Due to the low computational cost, the OLFC control technique proves to be a viable alternative for embedded systems of the AFM type. Works such as [34] make a comparison of computational costs.

Acknowledgements

The authors acknowledge the CAPES and CNPq, both Brazilian research funding agencies.

Conflict of interest

The authors declare no competing interests.

Author details

Mauricio A. Ribeiro[1]*, Jose M. Balthazar[1,2], Ângelo M. Tusset[1], Átila M. Bueno[3] and Hilson H. Daum[4]

1 Federal Technological University of Paraná—UTFPR, Ponta Grossa, PR, Brazil

2 School of Engineering of São Paulo State University—UNESP, Bauru, SP, Brazil

3 Polytechnic School of the University of São Paulo, São Paulo, SP, Brazil

4 Federal Technological University of Paraná—UTFPR, Guarapuava, PR, Brazil

*Address all correspondence to: mau.ap.ribeiro@gmail.com

References

[1] Bishop RH. The Mechatronics Handbook: 2 Volume Set. Boca Raton, Florida: CRC Press; 2002

[2] Younis MI. MEMS Linear and Nonlinear Statics and Dynamics. New York: Springer; 2011

[3] Balthazar JM, Tusset AM, Bueno AM, de Pontes Junior BR. On an Overview of Nonlinear and Chaotic Behavior and Their Controls of an Atomic Force Microscopy (AFM) Vibrating Problem. In: Awrejcewicz J, Hagedorn P, editors. Nonlinearity, Bifurcation and Chaos - Theory and Applications [Internet]. London: IntechOpen; 2012 [Cited: 24 Nov 2022]. Available from: https://www.intechopen.com/chapters/40431. DOI: 10.5772/51834

[4] Bueno AM, Balthazar JM, Piqueira JRC. Phase-Locked Loop design applied to frequency-modulated atomic force microscope. Communications in Nonlinear Science and Numerical Simulation. 2011;**16**(9):3835-3843

[5] Bueno AM, Balthazar JM, Piqueira JRC. Phase-locked loops lock-in range in frequency modulated-atomic force microscope nonlinear control system. Communications in Nonlinear Science and Numerical Simulation. 2012;**17**(7): 3101-3111

[6] Sajjadi M, Pishkenari HN, Vossoughi G. Dynamic modeling of trolling mode AFM: The effects of cantilever torsion, nanoneedle flexibility, and liquid nanoneedle interactions are considered. Ultramicroscopy. 2017;**182**:99-111

[7] Balthazar JM, Tusset AM. On dynamics, control and applications to engineering and life science-II. Mathematics in Engineering, Science and Aerospace: The Transdisciplinary International Journal. 2019;**10**:1-4

[8] Binnig G, Quate CF, Gerber C. Atomic Force Microscopy. Physical Review Letters. 1986;**56**:930. DOI: 10.1103/PhysRevLett.56.930

[9] Ribeiro MA, Kurina GA, Tusset AM, Balthazar JM. Nonlinear numerical analysis and averaging method applied atomic force microscopy with viscoelastic term. Achieves in Applied Mechanics. 2022;**92**:3817-3827. DOI: 10.1007/s00419-022-02264-5

[10] Hacker E, Gottlieb O. Local and global bifurcations in magnetic resonance force microscopy. Nonlinear Dynamics. 2020;**99**:201-225. DOI: 10.1007/s11071-019-05401-

[11] Ribeiro MA, Tusset AM, Balthazar JM. On nonlinear fractional dynamics analysis of atomic force microscopy system in trolling mode with viscoelasticity term. Brazilian Journal of Physics. 2022;**52**:186. DOI: 10.1007/s13538-022-01188-3

[12] Mohammadi SZ, Moghadam M, Pishkenari HN. Dynamical modeling of manipulation process in trolling-mode AFM. Ultramicroscopy. 2019;**197**:83-94

[13] Settimi V, Rega G. Local versus global dynamics and control of an AFM model in a safety perspective. In: Lenci S, Rega G, editors. Global Nonlinear Dynamics for Engineering Design and System Safety. CISM International Centre for Mechanical Sciences. Cham: Springer; 2019

[14] Haghighi MS, Sajjadi M, Pishkenari HN. Model-based topography estimation in trolling mode atomic force microscopy. Applied Mathematical Modelling. 2020;**77**:1025-1040

[15] Belardinelli P, Wiebe R, Alijani F, Lenci S. Machine learning to probe modal interaction in dynamic atomic force microscopy. Mechanical Systems and Signal Processing. 2022;**179**:109312

[16] Bradbury J, Merity S, Xiong C, Socher R. Quasi-recurrent neural networks. arXiv preprint arXiv: 1611.01576. 2016

[17] Ribeiro MA, Balthazar JM, Lenz WB, Rocha RT, Tusset AM. Numerical exploratory analysis of dynamics and control of an atomic force microscopy in tapping mode with fractional order. Skock and Vibrtion. 2020;**14**:4048307

[18] Ribeiro MA, Tusset AM, Lenz WB, et al. Numerical analysis of fractional dynamical behavior of Atomic Force Microscopy. European Physical Journal. 2021;**230**:3655-3661

[19] Tusset AM, Balthazar JM, Ribeiro MA, et al. Chaos control of an atomic force microscopy model in fractional order. European Physical Journal. 2021; **230**:3643-3654

[20] Tusset AM, Ribeiro MA, Lenz WB, et al. Time delayed feedback control applied in an atomic force microscopy (AFM) model in fractional-order. Journal of Vibrational Engineering Technology. 2020;**8**:327-335

[21] Yamaguchi PS, Tusset AM, Ribeiro MA, Balthazar JM. An Atomic Force Microscopy (AFM) Modelling in Fractional Order: Nonlinear Control System Design. Brazilian Journal of Physics. 2022;**52**:151. DOI: 10.1007/ s13538-022-01155-y

[22] Daza A, Wagemakers A, Georgeot B, Guéry-Odelin D, Sanjuán MA. Basin entropy: A new tool to analyze uncertainty in dynamical systems. Scientific Reports. 2016;**6**(1):1-10

[23] Daza A, Wagemakers A, Sanjuán MA. Ascertaining when a basin is Wada: The merging method. Scientific Reports. 2018;**8**(1):1-8

[24] McDonald SW, Grebogi C, Ott E, Yorke JA. Fractal basin boundaries. Physica D: Nonlinear Phenomena. 1985; **17**(2):125-153

[25] Aguirre J, Viana RL, Sanjuán MA. Fractal structures in nonlinear dynamics. Reviews of Modern Physics. 2009;**81**(1):333

[26] Grebogi C, McDonald SW, Ott E, Yorke JA. Final state sensitivity: An obstruction to predictability. Physics Letters A. 1983;**99**(9):415-418

[27] Wolf A, Swift JB, Swinney HL, Vastano JA. Determining Lyapunov exponents from a time series. Physica D: Nonlinear Phenomena. 1985;**16**(3): 285-317

[28] Rützel S, Lee SI, Raman A. Nonlinear dynamics of atomic–force–microscope probes driven in Lennard–Jones potentials. Proceedings of the Royal Society of London. Series A: Mathematical, Physical and Engineering Sciences. 2003;**459**(2036):1925-1948

[29] Rafukov M, Balthazar JM. On control and synchronization in chaotic and hyperchaotic systems. Communications in Nonlinear Science and Numerical Simulation. 2008;**13**:1246-1255

[30] Tusset AM. Controle Ótimo Aplicado em Modelo de Suspensão Veicular Não Linear Controlada através de amortecedor Magneto-Relógio. Porto Alegre; 2008

[31] Kirk RE. Optimal Control Theory: An Introduction. Dover, New York: Courier Corporation; 2004

[32] Ogata K. Engenharia de controle moderno: Tradução de Heloísa Coimbra de Souza. 4th ed. SP, Brasil: Prentice-Hall; 2003

[33] Shawky AMEA. Position control of flexible manipulator using non-linear H∞ with state-dependent Riccati equation. Journal of Systems and Control Engineering. 2007;**2007**:475-486

[34] Daum HH, Tusset AM, Ribeiro MA, Litak G, Bueno AM, Balthazar JM. Analysis and chaos control of a four-dimensional magnetohydrodynamic model with hyperchaotic solutions. European Physical Journal-Special Topics. 2021;**1**:1-11

[35] Tusset AM, Janzen FC, Rocha RT, Balthazar JM. On an optimal control applied in MEMS oscillator with chaotic behavior including fractional order. Complexity. 2018;**1–12**:2018

[36] Daum HH, Tusset AM, Ribeiro MA, Balthazar JM, Bueno AM, Litak G. Dynamics and control of a vibrating system with hyperchaotic behavior using an electronic circuit implementation. Brazilian Journal of Physics. 2022;**52**:104